The Proceedings of the International Conference on Electrical Systems & Automation

Mohamed Bendaoud · Borutzky Wolfgang ·
Amine El Fathi
Editors

The Proceedings of the International Conference on Electrical Systems & Automation

Recent Advances in Renewable Energy—Volume 1

Springer

Editors
Mohamed Bendaoud
ENSA Khouribga, Sultan Moulay Slimane University
Khouribga, Morocco

Amine El Fathi
FST, AL-Hoceima, Abdelmalek Essadi University
AL-Hoceima, Morocco

Borutzky Wolfgang
Bonn-Rhein-Sieg University of Applied Sciences
Sankt Augustin, Germany

ISBN 978-981-19-0034-1 ISBN 978-981-19-0035-8 (eBook)
https://doi.org/10.1007/978-981-19-0035-8

This Springer imprint is published by the registered company Springer Nature Singapore Pte Ltd.
The registered company address is: 152 Beach Road, #21-01/04 Gateway East, Singapore 189721, Singapore

Committees

Honorary Committee

Hmina Nabil, President of Sultan Moulay Slimane University, Morocco
Sajieddine Mohammed, Director of ENSA-Khouribga

General Chairs

Mohamed Bendaoud, Sultan Moulay Slimane University, Morocco
Chikh Khalid, Sultan Moulay Slimane University, Morocco

Conference Co-chairs

Elbarbri Noureddine, Sultan Moulay Slimane University, Morocco
Borutzky Wolfgang, Bonn-Rhein-Sieg University of Applied Sciences, Germany
Ailane Abdellah, Sultan Moulay Slimane University, Morocco

Publication Chairs

Mohamed Bendaoud, Sultan Moulay Slimane University, Morocco
Borutzky Wolfgang, Bonn-Rhein-Sieg University of Applied Sciences, Germany
Chikh Khalid, Sultan Moulay Slimane University, Morocco
Amine El Fathi, Abdelmalek Essadi University, Morocco

Sponsorship Co-chairs

Benchagra Mohamed, Sultan Moulay Slimane University, Morocco
Khamlich Salahddine, Sultan Moulay Slimane University, Morocco
Amharech Amine, Sultan Moulay Slimane University, Morocco

Communication Co-chairs

Massour El Aoud Mohamed, Sultan Moulay Slimane University, Morocco
Rhofir Karim, Sultan Moulay Slimane University, Morocco
Maaider Kamal, Sultan Moulay Slimane University, Morocco

Technical Programme Chairs

Hihi Hicham, Sidi Mohamed Ben Abdellah University, Morocco
Lokriti Abdesslam, Sultan Moulay Slimane University, Morocco
Er Rachid Ismail, Sultan Moulay Slimane University, Morocco
Amer Ragab Zerek, Zawia University, Libya

Technical Programme Committee

Abouzahir Mohamed, Mohamed V University, Morocco
Abouobaida Hassan, Chouaib-Doukkali University, Morocco
Afia Aziz, Hassan II University, Morocco
AlFaouri Yanal Shaher, University of Jordan, Jordan
Amer Ragab Zerek, Zawia University, Libya
Asmaa A. Abdeewi, Polymer Research Center, Lybia
Ashok Suryawanshi, Pimpri Chinchwad College of Engineering, India
Barbosa Karina, University of Santiago, Chile
Belkacem Ould bouamama, Polytech of Lille, France.
Binti Mohamed Shah Noraisyah, Universiti Malaya, Malaysia
Boulghasoul Zakaria, Cadi Ayyad University, Morocco
Caines Peter, McGill University, Canada
Chuah joon huang, Universiti Malaya, Malaysia
Elbacha Abdelhadi, Cadi Ayyad University, Morocco
El Aamri Faicel, Hassan II University, Morocco
EL bahir Lhoussain, Cadi Ayyad University, Morocco
El akchioui Nabil, Abdelmalek Essadi University, Morocco

Preface

This edited volume on "Recent Advances in Renewable Energy" is an outcome of the selected papers presented at the International Conference on Electrical Systems & Automation, (ICESA'21) which was held on 8–9 December 2021 at the National School of Applied Sciences Khouribga, Morocco. This conference brings together researchers in industry and academia to exchange their ideas, applications and innovative techniques in the field of renewable energy and electrical systems.

The book provides rigorous discussions, the state of the art and recent developments in the field of renewable energy sources supported by examples and case studies, making it an educational tool for relevant undergraduate and graduate courses. The book will be a valuable reference for beginners, researchers and professionals interested in renewable energy.

The conference programme featured a rich set of keynote session topics and tutorials that extend beyond the papers contained in these proceedings. There were six high-profile keynotes by eminent researchers: Prof. Yazami Rachid (KVI Pte Ltd., Singapore), Prof. Hachimi Hanaa (Sultan Moulay Slimane University, Morocco), Prof. Pierluigi Siano (University of Salerno, Italy), Prof. Michael Schmidt (Offenburg University, Germany) and Prof. Yousfi Driss (Mohammed I University, Morocco).

The editors express their gratitude to Springer Nature authority for publishing this volume in the Springer Conference Proceedings Series.

The editors also express sincere thanks to the reviewers for their dedication in reviewing the articles. Also, thanks to the authors for submitting their articles to this volume.

It is hoped that this volume will be a good reference manual for researchers and budding engineers.

Khouribga, Morocco — Mohamed Bendaoud
Sankt Augustin, Germany — Borutzky Wolfgang
AL-Hoceima, Morocco — Amine El Fathi

Contents

About the Editors

Prof. Mohamed Bendaoud received the M.Sc. and Ph.D. degrees in Electrical Engineering from Cadi Ayyad University, Marrakech, Morocco, in 2012 and 2019, respectively. He is currently an assistant professor at National School of Applied Sciences (ENSA) of Khouribga, Sultan Moulay Slimane University-Morocco.

He is the founder and the general chair of the International Conference on Electrical Systems & Automation. He has served and continues to serve on the technical programme committees and as a reviewer of numerous international conferences and journals such as *Journal of The Franklin Institute*, *Mechatronic Systems and Control*. He is the editor-in-chief of the new ambitious *Journal of Electrical Systems & Automation* (JESA). He is the deputy director of the science and technology laboratory for engineers.

His research interests include:

- Modelling and control of grid connection for photovoltaic and wind energy.
- Modelling and simulation methodologies for multidisciplinary systems, in particular bond graph based.
- Control of power converters.

Prof. Borutzky Wolfgang is a retired professor for Modelling and Simulation of Engineering Systems at Bonn-Rhein-Sieg University of Applied Sciences, Germany. He obtained his University Diploma Degree in Mathematics in 1979 and his Doctoral Degree in Mechanical Engineering in 1985, both from the Technical University of Braunschweig, Germany.

He was a visiting professor at universities in the Netherlands, the USA, France and in Argentina. Since 2008, he holds an honorary position as an associate professor of Electrical Engineering and Information Technology at the University of Dubrovnik, Croatia.

His main research interests include modelling and simulation methodologies for multidisciplinary systems, in particular bond graph based as well as object-oriented modelling; modelling, analysis, control and fault diagnosis of mechatronic systems and hybrid failure prognostic methods; modelling languages; scientific computing and numerical algorithms for continuous system simulation.

Dr. Borutzky has published in the proceedings of many peer-reviewed international conferences on modelling and simulation and in refereed scientific journals. In 2019, he received the best paper award of the 12th International Conference on Integrated Modelling and Analysis in Applied Control and Automation (IMAACA 2019) in Lisbon, Portugal, which is part of the annual International Multidisciplinary Modelling & Simulation Multiconference (I3M).

He is the author of a 2010 Springer monograph on bond graph methodology and of a 2015 Springer book on bond graph model-based fault diagnosis in hybrid systems. He is also the editor and a co-author of a 2011 Springer compilation text on Bond Graph Modelling of Engineering Systems and of a 2016 compilation text entitled Bond Graphs for Modelling, Control and Fault Diagnosis with contributions from experts in various fields from all over the world. His latest book published with Springer is titled Bond Graph Modelling for Control, Fault Diagnosis and Failure Prognosis. Furthermore, he was the guest editor of two special journal issues on bond graph modelling.

Since 1990, he has served in many international scientific conferences on modelling and simulation in various capacities. In 2005, he was an assistant general chair of the European Conference on Modelling and Simulation (EMCS) in Riga, Latvia, in 2006 he was the general chair of the ECMS 2006 in Sankt Augustin, Germany, and in 2013, he was the conference chair of the International IASTED Conference on Modelling, Identification, and Control (MIC 2013) in Innsbruck, Austria. From 2005 to 2007, he served on the Board of the European Council for Modelling and Simulation.

In 2009, he was an invited tutorial speaker of the IASTED Conference on Modelling, Simulation and Identification (MSI 2009) in Beijing, China, and in 2010, he was one of the invited keynote speakers to the Conference on Power Control and Optimisation (PCO 2010) in Kuching, Malaysia.

In 2012, he was one of the invited keynote speakers of a workshop on bond graph modelling in Pune, India, and one of the invited keynote speakers of the SCS Summer Simulation MultiConference in Genoa, Italy.

In 2013, he was an invited tutorial speaker of the IASTED Conference on Control and Applications (CA 2103) in Honolulu, USA, and one of the invited keynote speakers of the International Conference on Smart Technologies for Mechanical Engineering at Delhi Technological University (DTU), Delhi, India.

In 2018, he was one of the invited keynote speakers of the 1st International Conference on New Frontiers in Engineering, Science & Technology (NFEST 2018), New Delhi, India, and of the 1st International Conference on Electronics, Control, Optimization and Computer Science (ICECOCS'18), Kenitra, Morocco.

He has also served as an invited external examiner and as a member of Ph.D. defence committees in France, in India, in the UK, and in Norway. Moreover, he has given invited lectures and short courses in industry on bond graph modelling.

During the 2004–2006 biennium, he served on the SCS Board of Directors. He was also active as a member of the Editorial Board of some major modelling and simulation-related journals. Since 2009, he is again an associate editor of the journal *Simulation: Transactions of the Society for Modelling and Simulation International.*

Prof. Amine El Fathi received the M.Sc. and Ph.D. degrees in Electrical Engineering & Renewable Energies from Cadi Ayyad University, Marrakech, Morocco, in 2012 and 2017, respectively. He is currently an assistant professor at the Faculty of Sciences and Techniques of Al Hoceima-Abdelmalek Essaâdi University,Morocco. He has served and continues to serve on technical programme committees of several international conferences. He served also as a reviewer in numerous highly ranked journals such as *Energy Conversion and Management*. He is one of the main instigators of the new ambitious *Journal of Electrical Systems & Automation* (JESA).

His current research interests include:

- Performance parameters assessment of stand-alone renewable energy systems.
- Evaluation of lead–acid batteries under real operating conditions.
- Economic assessment of renewable energy systems (levelized cost).
- Modelling and control of grid connection for photovoltaic and wind energy.

MPPT-Based a Fuzzy Logic and PO Algorithm for Standalone PV System Under Partial Shading Conditions

Farid Oufqir, Mohamed Bendaoud, and Khalid Chikh

Abstract Due to the importance of renewable energy applications, which has gotten a lot of attention throughout the world, the use of standalone photovoltaic systems is becoming more prominent. However, to get the optimal performance under various operating conditions, an efficient design control must be performed. In this study, a two-stage single-phase standalone PV system with battery storage is successfully developed. The first stage employs a DC-DC boost converter based on an MPPT controller. A DC-AC inverter along with LC filter is used in the second stage to supply the load. The system has been modelled and simulated under uniform and partial shading conditions (PSC) to extract the maximum power point. The fuzzy logic controller was designed and compared to the conventional perturb and observe (P&O).

Keywords Photovoltaic · Converter control · Partial shading conditions · Fuzzy logic MPPT · Standalone system

Nomenclature

I_{ph}	Current of PV cell
I_d	Diode current
I_{sh}	Current in shunt resistance R_{sh}
R_s	Serie resistance
R_{sh}	Parallel resistance
V_{th}	Thermal potential
k	the Boltzmann constant ($1381 \cdot 10^{-23}$ J/K)

F. Oufqir (✉) · M. Bendaoud · K. Chikh
Science and Technology for the Engineer Laboratory (LaSTI), ENSA Khouribga, Khouribga, Morocco
e-mail: farid.oufqir@usms.ac.ma

M. Bendaoud
e-mail: m.bendaoud@usms.ma

M. Bendaoud et al. (eds.), *The Proceedings of the International Conference on Electrical Systems & Automation*, https://doi.org/10.1007/978-981-19-0035-8_1

Ns	Number of series-connected cells
q	Electron charge ($1.6 \cdot 10^{-19}$)
I_o	Reverse saturation current
A	Diode ideality factor
I_{sh}	Current in parallel resistance
Icc	Light generated current at nominal operating condition [A]
K_i	Temperature coefficient
T_n	Nominal temperature
T	Actual temperature
G	Irradiance of the surface
G_0	Nominal irradiance
V_o	DC bus voltage
I_o	DC bus current
α	Duty cycle of the boost converter
F	Switching frequency of the boost converter
Δ_V	Voltage ripple
$\alpha 1$	Duty cycle of the bidirectional converter in buck mode
$\alpha 2$	Duty cycle of the bidirectional converter in boost mode
V_{bat}	Battery voltage
F1	Switching frequency of the bidirectional converter
fc	Resonant frequency
Fi	Switching frequency of the inverter
P(k)	Power of the GPV at sampling time k
V(k)	Voltage of the GPV at sampling time k

1 Introduction

With the growing energy demand, the evolution of technology has enabled an important development of renewable energies [1]. These clean and sustainable energies, particularly photovoltaic (PV) energy, have grown in prominence as an alternative to fossil fuels. the PV power has become a reliable power source over the last decade, it makes it possible to become energy self-sufficient. With the technical advancements that occur every year in the industry, it is expected that solar PV systems will develop and become more accessible to as many people as possible [2].

In this paper, we are particularly interested in standalone PV system, with a battery storage that stabilizes the DC bus voltage and makes easy the connection between PV modules, batteries, and AC load. the storage battery is necessary to maintain a constant supply of electrical energy, and any extra power produced may be accumulated for later use [3]. Battery charge and discharge control is provided by the bidirectional converter. Another DC/DC converter is employed, it is a boost converter that transforms the DC voltage of the panels to a higher voltage and is driven by a Maximum power point tracking (MPPT) controller that is designed to deal with the constraints caused by various climate variations. This controller's performance is determined by how fast it achieves the maximum power point (MPP), how to oscillate around that point, and how robust this controller is to unexpected variations such as partial shading condition (PSC) [4].

Under partial shading conditions, varying radiation on each PV array generates different power from one PV module to another. Furthermore, multiple peaks will be created because of the use of bypass diodes to protect the shaded PV modules from hot spot points [5]. As a result, tracking the global maximum power is essential for achieving high power efficiency. To reduce power loss, various PV array configurations such as honeycomb, series–parallel, bridge connected, zigzag, sudoku, can be used [6, 7].

Currently, the MPPT methods have been the subject of many studies to improve the dynamic performance of the PV system, namely the capacity to quickly follow the global maximum power point (GMPP) in the presence of various local maximum power points (LMPP) during the PSC. Conventional MPPT techniques, especially incremental conductance, constant voltage and perturb and observe (P&O), are effective, and reliable in tracking the specific peak under uniform conditions. although, if the GMPP was not at the first peak, they failed to track it under PSC and stick at LMPP [8]. However, Particle swarm optimization (PSO) [9] and grey wolf optimizer (GWO) [10] are bio-inspired heuristic MPPT algorithms that can follow the GMPP.

In [7] a hybrid technique combining the incremental conductance method and the grey wolf algorithm (GWA) is presented. The findings demonstrate that the suggested hybrid technique tracks the GMPP in the scenarios of uniform solar irradiation and partial shading effects.

Kermadi et al. [11] presents a hybrid of adaptive P&O and particle swarm optimization (PSO). A PV array simulator-based experimental prototype is used. It was observed that the proposed approach converges to GMPP, and that tracking is assured even in complex partial shadow conditions, however, it struggles from drift under quickly changing irradiance values. Using fuzzy logic (FL) and a dual (MPPT) controller, [12] provides a new digital control technique for a photovoltaic (PV) system. It employs a novel perturb and observe (P&O) technique based on a fuzzy-logic controller (FLC) to keep the system power operating point at its maximum. The suggested control approach ensures that the system operates in a stable manner.

Oufqir et al. [13] presents a model for a standalone PV system. The results demonstrate that the suggested system maintains a steady frequency and RMS load voltage while minimizing THD, but the MPPT has not been checked under PSC.

The authors of this paper present a design for a standalone PV system (Fig. 1), with a series–parallel configuration for solar panels. The major contribution of this work is to elaborate a comparative analysis to highlight the performances of the P&0 and FL. The methodology will be helpful for the system designers.

This paper is structured as follows: Section 2 contains a detailed system description as well as mathematical modeling of the PV panel. Sections 3 and 4 gives more details about control strategies with converters sizing, energy storage and MPPT algorithms. Section 5 shows the simulation results using MATLAB/Simulink to check the

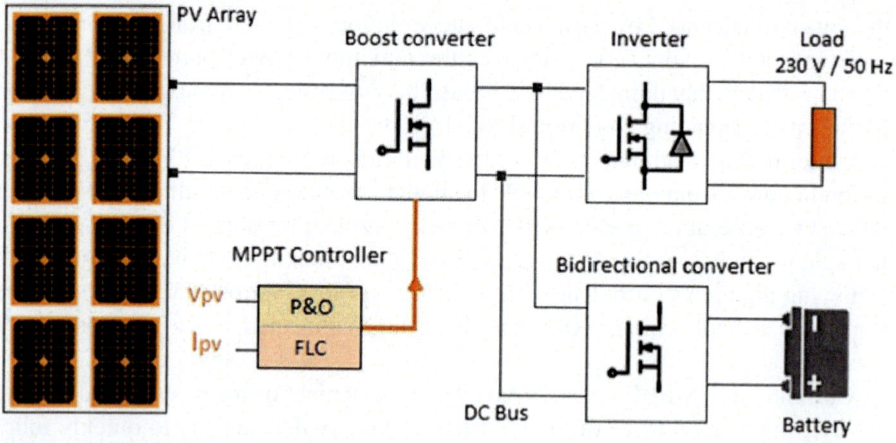

Fig. 1 Structure of the standalone PV system

robustness of the control strategy, and the validation of the proposed MPPT method under PS conditions.

2 System Description

Figure 1 represents the synoptic diagram for the system simulated under MATLAB/Simulink. This system consists of a 2 kW to 1000 W/m^2 PV generator which is coupled to a load through a Boost converter, a single-phase inverter and a bidirectional converter that plays a key role in stabilizing the DC bus voltage. In this work, we chose the Boost converter because we have a maximum voltage of 120 V and a current of 16.6 A that the PV generator can deliver to the input of the boost converter, and we need a voltage of 400 V at the input of the inverter. The MPPT controller is implemented in the Boost converter to extract the maximum power from the PV generator. This PV generator consists of 2 PV strings interconnected in parallel. Each PV string contains 4 PV modules interconnected in series. The PV model parameters used are given in Table 1.

Table 1 Parameters of PV panel

Parameters	Value
Maximum power	249 W
Open circuit voltage	36.8 V
Short circuit current	8.83 A
The voltage at MPP	30 V
The Current at MPP	8.3 A

2.1 Mathematical Modeling of PV Module

The PV cell can be represented by a single diode model (Fig. 2). This model consists of a current generator, corresponding to the photo-generated current, two resistors, and a diode. The mathematical model of the PV cell is given by [13]:

$$I_{pv} = I_{ph} - I_d - I_{sh} \tag{1}$$

The equation of I_d current can be represented as:

$$I_d = I_0\left(Exp\left[\frac{V + (I \times R_S)}{A \times V_{th}}\right] - 1\right) \tag{2}$$

The thermal potential (V_{th}) is given by:

$$V_{th} = \frac{K \times T \times N_S}{q} \tag{3}$$

The current I_{sh} in parallel resistance is described as:

$$I_{sh} = \frac{R_S.I_{pv} + V}{Rsh} \tag{4}$$

The current of the PV cell is proportional to the irradiance (G):

$$I_{Ph} = [I_{CC} + K_i(T - T_n)] * \frac{G}{G_0} \tag{5}$$

Figure 3 illustrates the I–V and P–V characteristics of the solar panel at various levels of irradiation while maintaining a constant temperature. Temperature and irradiance influence these properties; as the irradiance rise, so does the power of the photovoltaic panel. As a result, a control system is necessary to guarantee that the PV system is running at its maximum power point [14].

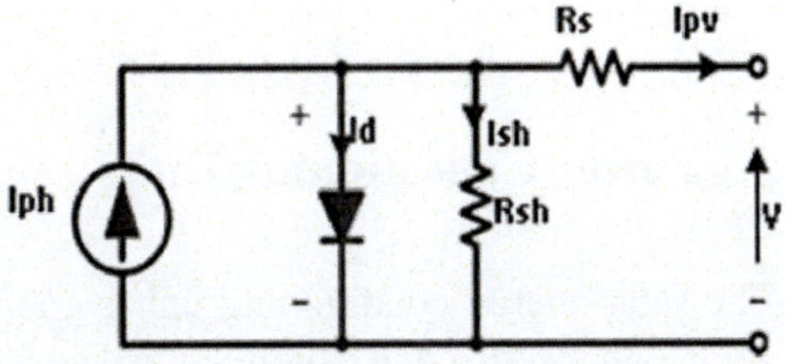

Fig. 2 Circuit diagram of PV cell

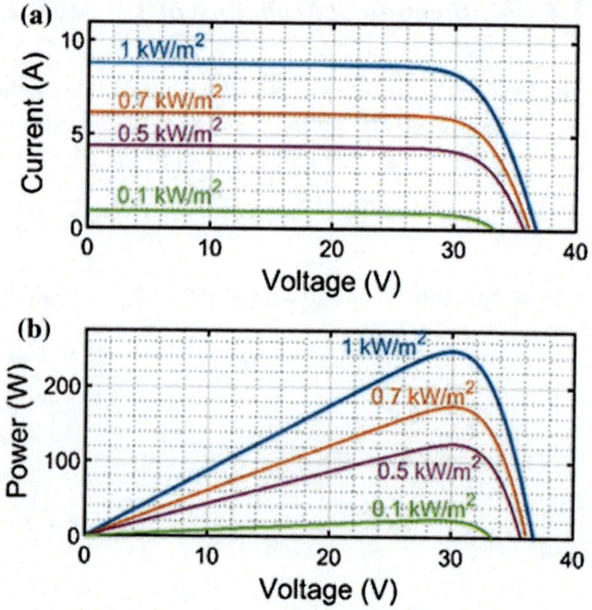

Fig. 3 PV characteristics for a temperature of 25 °C, **a** I–V curve and **b** P–V curve

2.2 *Partial Shading Effect*

According to [15], partial shading has a significant effect on PV output energy. When one part of the PV panel receives less radiation than another, the current produced by the lighted cells is higher than the current produced by the shaded cells, causing the shaded cells' diode to be reverse biased. As a result, power will be lost and potentially resulting in a hot spot problem with irreversible damage to the PV panel. So, the bypass diodes can be connected in parallel with PV cells to solve this issue [16].

Two strings of four PV panels were connected in parallel to simulate the effect of PSC (Fig. 4b). Four patterns were considered in this analysis, the data of which are presented in Table 2. In the first case, the irradiation of each module is equal to 1 kW/m^2, which is known as the standard test condition (STC), and various PS conditions are considered in patterns 1, 2 and 3.

3 Power Conditioning Units

The solar energy conditioning unit comprises a boost converter controlled via an MPPT, an inverter, and a bidirectional converter. It extracts the power available from the solar panels and stores it in the battery bank according to the state of the load to be supplied. Due to the constant use of energy, if the supplied power drops below the

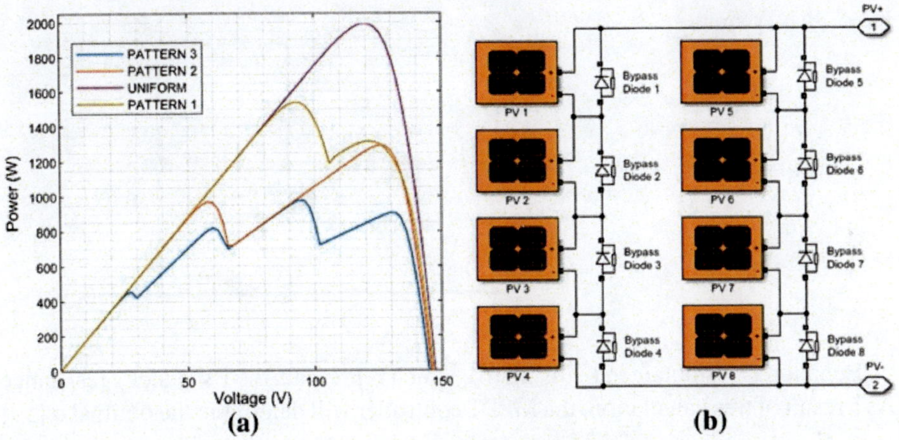

Fig. 4 **a** Power vs voltage plot under various shading patterns. **b** PV array arrangement

Table 2 Shading patterns

Patterns	Irradiauce levels (kW/m^2) for each PV panel					Position of GMPP peak	Power at GMPP (KW)
	PV 1	PV 2&6	PV 3&7	PV 4&8	PV 5		
Uniform	1	1	1	1	1	1st	1.992
Pattern 1	0.3	1	1	1	1	1st	1.538
Pattern 2	1	1	0.6	0.6	1	2nd	1.292
Pattern 3	1	0.8	0.6	0.4	1	3rd	0.976

level requested by the load, the battery bank will discharge automatically to transfer the energy and meet the load requirements [17].

3.1 Boost Converter

The boost converter allows the load and the PV generator to be adapted, to draw the most energy. Depending on the state of the switch (S), the functioning of a Boost converter can be separated into two distinct stages. When the switch is closed, the current in the inductor increases, resulting in the storage of energy in the form of magnetic energy. When the switch is turned on, the inductor is connected to the generator, causing the energy collected in the inductor to be transferred to the capacitor. The boost converter circuit is depicted in Fig. 5.

The output voltage (V_o), and the output current (Io) can be defined by the following equations [18]:

$$I_o = I(1-\alpha),\ V_o = \frac{1}{1-\alpha}V \tag{6}$$

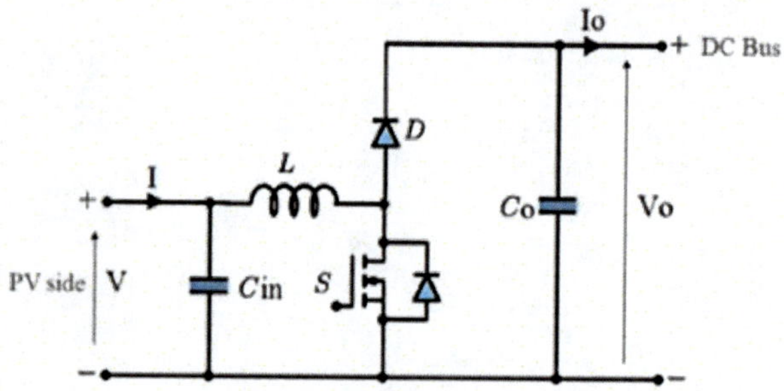

Fig. 5 Boost converter

Equation (7) is obtained by using (6), which represents the PV panel's resistance. As a result of this calculation, the MPPT controller will determine the optimal α [19].

$$R_{eq} = \frac{V}{I} = \frac{V_o}{I_o}(1-\alpha)^2 = R(1-\alpha)^2 \tag{7}$$

The inductor selection has a significant impact on the boost converter's efficiency. The inductance value can be given as follows, where F is the switching frequency:

$$L_{min} = \frac{\alpha(1-\alpha)^2 V_O}{2FI_O} \tag{8}$$

An input capacitor is used to reduce the ripple in the input voltage and to provide an alternative current to the inductor. the value can be calculated by (9), where Δ_V is the voltage ripple:

$$Cin_{min} = \frac{\alpha V}{8F^2 L\Delta V} \tag{9}$$

The value of the output capacitor may be determined as follows (10), by using output voltage ripple ΔV_O:

$$\mathrm{Co_{min}} = \frac{\alpha \mathrm{I_o}}{\mathrm{F}\Delta\mathrm{Vo}} \tag{10}$$

3.2 *Bidirectional Buck-Boost Converter*

The storage battery allows the energy produced by the solar panel to be stored for subsequent use. The battery with the bidirectional converter is utilized to maintain a constant DC bus voltage and distribute the energy to the load when there is a power outage [20].

At the outset, it is important to define the energy storage and photovoltaic power supply system specifications. The DC bus has a rated voltage (V_o) of 400 V. The storage device consists of a series connection set of lead-acid batteries with 220 Ah total capacity and rated voltage 48 V. Nominal parameters of the battery are shown in Table 3.

This converter (Fig. 6) works in two modes (buck and boost), and according to these modes, we determine the design of its elements. The selection of the inductor is a critical design factor in a high-power converter [21].

When the converter works in the Buck mode, the inductor value can be expressed as:

$$L_{\text{buck}} = \frac{\alpha_1(1-\alpha_1)V_O}{2F\Delta Ib} \tag{11}$$

α_1 is the duty cycle in the buck mode: $\alpha_1 = \frac{V_{bat}}{V_O}$

In the boost mode, the expression is:

$$L_{\text{boost}} = \frac{\alpha_2 V_{bat}}{2F\Delta Ib} \tag{12}$$

Table 3 Battery bank parameters

Lead-acid battery parameters	Value
Rated capacity	220 Ah
Nominal voltage	12 V
Number of parallel connections	1
Number of series connections	4
Total nominal voltage	48 V

Fig. 6 Bidirectional converter

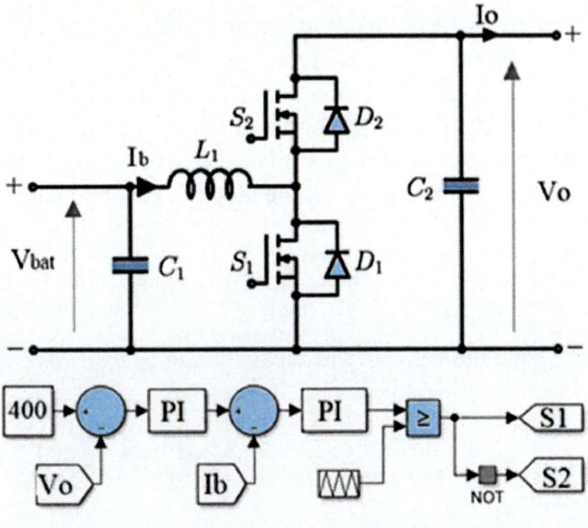

α_2 is the duty cycle in the boost mode: $\alpha_2 = 1 - \frac{V_{bat}}{V_O}$

Therefore, the value of the converter inductor should be the minimum of the value given by L_{buck} and L_{boost} :

$$L_1 = \text{Min}[L_{\text{buck}},\ L_{\text{boost}}]$$

Also, the output and input capacitor values are determined by the capacitor's voltage ripple as shown below (13):

$$C_1 = \frac{\alpha_1(1-\alpha_1)V_O}{8{F_1}^2 L\Delta V_{bat}} \tag{13}$$

$$C_2 = \frac{\alpha_2 I_O}{F_1 \Delta V_O} \tag{14}$$

The converter's controller will reduce the ripple of the current and voltage in the input or output, as well as increase the converter's efficiency. The reliability of a bidirectional DC/DC converter cannot be achieved by a single closed loop. As seen in Fig. 6, we use a voltage and current double closed-loop [22].

The control circuit consists of the outer voltage loop and the inner current loop. In the outer voltage loop, we calculate V_o and compare it to Vo_{ref} (400 V). The gap of two is transmitted to the PI regulator, then move to the current loop. Finally, as compared to *Ib*, the difference through by the other PI regulator into the PWM modulation. The controller will obtain a steady voltage in V_o and a stable current in Ib.

3.3 *Inverter Circuit*

The AC load is linked to the system by a single-phase inverter, which allows the controller in Fig. 7 to transform direct current to alternating current. To suppress high frequency harmonics, an LC filter is employed at the inverter's output.

The LC filter is a second order filter, the capacitor serves as the shunt element and must be chosen in such a way as to have a low reactance at the switching frequency (F_i) [23]. The resonant frequency (f_C) is obtained from (15):

$$\mathrm{f_C} = \frac{1}{2\pi}\frac{\mathrm{V_O}}{\sqrt{\mathrm{L_2C_4}}} \text{ and } \mathrm{f_C} \le \frac{\mathrm{F_i}}{10} \tag{15}$$

Then the value of the capacitor can be expressed as (16):

$$\mathrm{C_4} = \frac{1}{\mathrm{L_2}}\left(\frac{10}{2\pi \mathrm{F_i}}\right)^2 \tag{16}$$

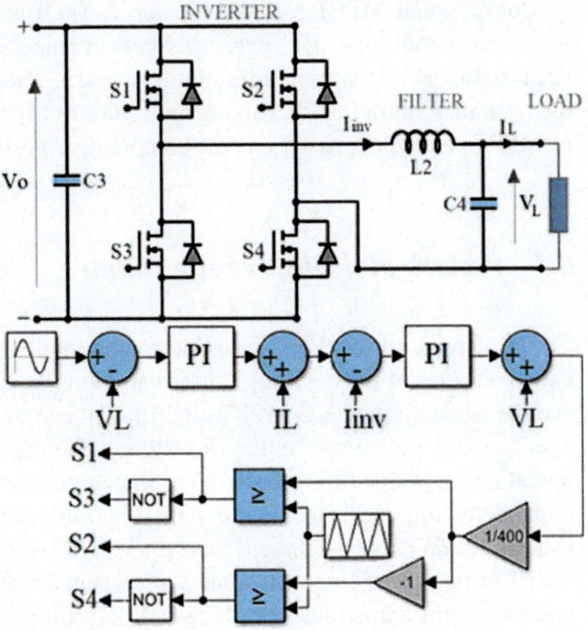

Fig. 7 Circuit of the single-phase inverter

The inductor eliminates the low frequency harmonic components. The ripple current (ΔI_{inv}) is influenced by the DC bus voltage, inductance, and switching frequency. Since the DC voltage (V_O) and switching frequency (F_i) are fixed, the inductance can be evaluated using Eq. (17):

$$L_2 = \frac{1}{8}\frac{V_O}{F_i \Delta I_{inv}} \tag{17}$$

The double closed-loop proportional-integral (PI) controller with internal current loop and external voltage loop is implemented to increase the efficiency of the inverter and the output voltage waveform quality. The inner loop is used to improve the system's dynamic performance and quickly eliminate the effects of load disturbance; the outer loop is used to improve the system's static performance [24].

4 MPPT Controller

The PV module's output power is normally intermittent. As a result, the DC-DC boost converter with an MPPT controller should be made to follow the MPP. Overall, the MPPT contributes primarily to the reduction of PV system costs and the enhancement of performance. Many approaches, including conventional and soft computing MPPT techniques, have been used.

Conventional MPPT techniques such as (PO) give good results under uniform irradiance conditions. However, this performance is not assured during the PSC because the P–V curve includes multiple peaks. Therefore, PO could be trapped by the local maximum LMPP. Among the solutions to overcome this problem is the use of MPPT controllers based on soft-computing algorithms [15].

4.1 *Perturb and Observe Algorithm*

The basic principle of (P&O) is to interrupt a small amplitude voltage and then evaluate the action of the resulting power variation [8]. Figure 8a depicts the flow chart and the model of the P&O MPPT algorithm. It can be deduced that if the power rises after a voltage disturbance, the direction of the disturbance is retained. If not, the operation is reversed to restart convergence to the new PPM. Figure 8b depicts the implementation of this technique in the Simulink environment. The findings in [25] show that P&O is less efficient than (IncCond) but outperforms other conventional MPPT techniques. Moreover, some authors found that some modifications on (P&O) placed it in the same rank with IncCond [26]. In general, conventional MPPT techniques are not exact or effective in dealing with the PS situation since they cannot follow the GP under PSC, while soft computing or heuristic techniques can track the GP and effectively deal with the PS problem.

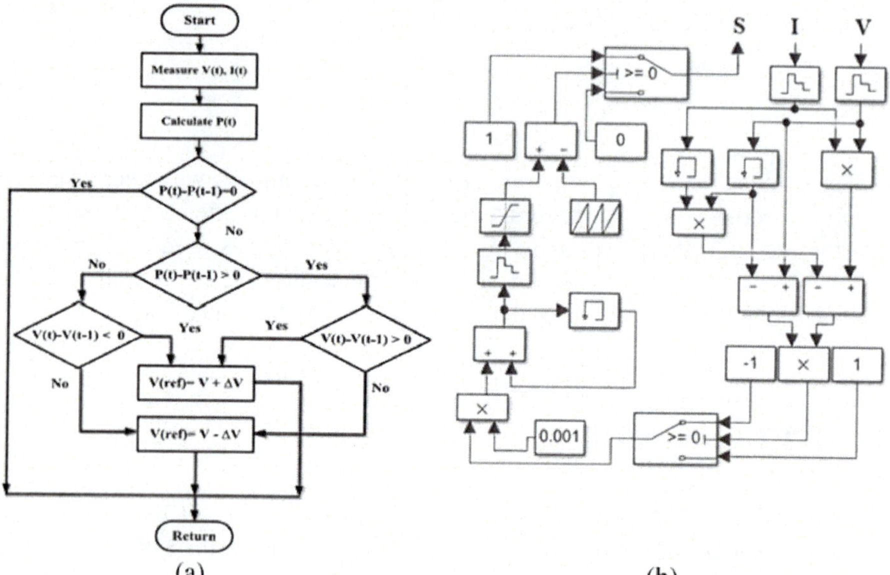

(a) (b)

Fig. 8 **a** Flow chart of conventional P&O and **b** Model of P&O controller in simulink

4.2 *Fuzzy Logic Controller*

The flowchart of the fuzzy controller is shown in Fig. 10. This MPPT employs fuzzy logic to determine the maximum power point (MPP). This control algorithm consists of three steps, namely:

(a) *Fuzzification*

Fuzzification allows real variables to be converted to fuzzy variables. The PV module's real voltage and current are continuously measured, and the power may then be evaluated. The control is determined based on the satisfaction of two criteria concerning two input variables, which are the error (E) and the error change (CE), at sampling time k. the scale factors KE, KCE and Kd will be changed based on the output performances [27].

The variables E and the CE are expressed as follows:

$$E(k) = \frac{P(k) - P(k-1)}{v(k) - v(k-1)} \tag{18}$$

$$CE(k) = E(k) - E(k-1) \tag{19}$$

where P(k) and V(k) are respectively the power and voltage of the GPV. Therefore, E(k) is zero at the MPP of the GPV. These input variables are expressed in terms of linguistic variables or labels such as PB (positive big), PS (positive small), ZE (zero), NS (negative small), NB (negative big). Figure 9 shows the corresponding input and output membership functions.

(b) *Inference method*

Table 4 [28] shows the rules of the fuzzy controller, where all the inputs to the matrix are the fuzzy sets of error (E), change in error (CE) and change in duty cycle (dD) converter. In the case of fuzzy control, the control rule must be designed so that the input variable E is always zero. In our case, we use the Mamdani inference method which is the MAX–MIN fuzzy combination.

(c) *Defuzzification*

Inference methods provide a function for the resulting membership variable, so this is fuzzy information. Given that the DC-DC converter requires an accurate control signal D at its input, this fuzzy information must be transformed into determined information. this transformation is called defuzzification.

5 Results and Discussion

Table 5 presents the system parameters utilized in the simulation.

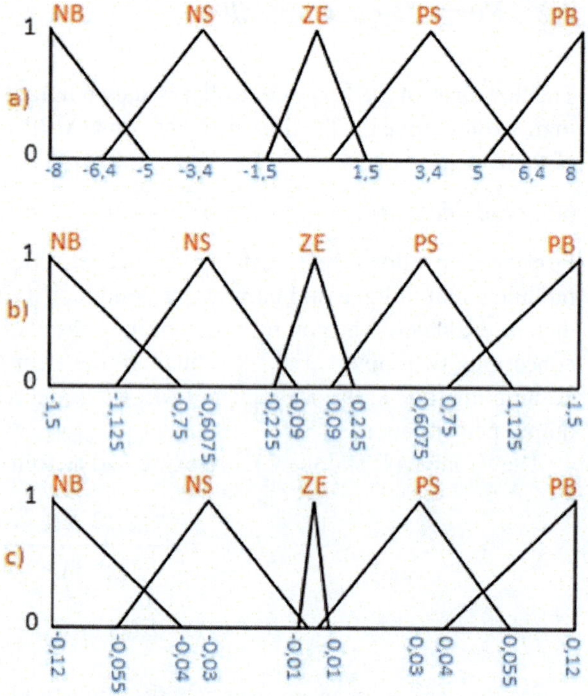

Fig. 9 **a** Membership function for (E), **b** Membership function for (CE), **c** Membership function for increment of duty ratio command

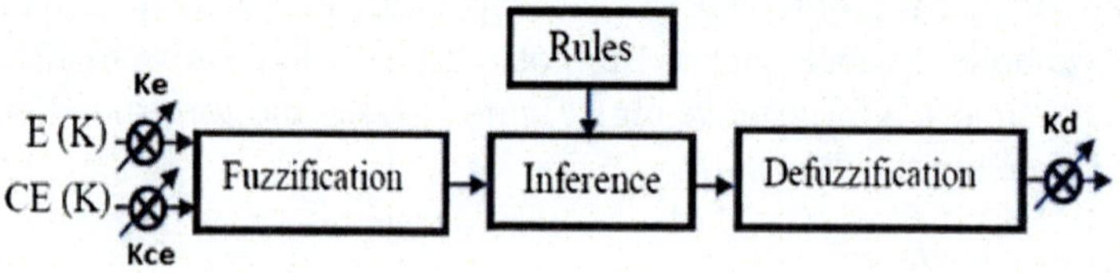

Fig. 10 Flowchart of the fuzzy controller

Table 4 Fuzzy rules database

Fuzzy rule		E(K)				
		NB	NS	ZE	PS	PB
CE(k)	NB	ZE	PB	PS	ZE	NB
	NS	PB	PS	ZE	ZE	NB
	ZE	PB	PS	ZE	NS	NB
	PS	PB	ZE	ZE	NS	NB
	PB	PB	ZE	NS	NB	ZE

Table 5 System parameters

Parameters	Values
Capacitance, C_{in}	500 μF
Capacitance, C_o	400 μF
Inductance, L	60 μH
Switching frequency—Boost converter	50 kHz
DC bus capacitor, C_3	500 μF
Capacitance, C_1	500 μF
Capacitance, C_2	50 μF
Inductance, L_1	80 μH
Switching frequency—bidirectional converter	20 kHz
Battery	220 Ah
Battery voltage	48 V
Inductance, L_2	4.06 mH
Capacitance, C_4	6.23 μF
Switching frequency—inverter	10 kHz
DC link reference voltage	400 V
AC voltage reference	230 V

5.1 System Under Uniform Irradiance

The first simulation's purpose is to validate the robustness of the suggested control technique in response to uniform irradiance at various levels. In this scenario, the power load is set to 2KW, and the irradiance goes from 700 to 1000 W/m^2. This scenario allows you to test the system's operation, specifically the SOC of the batteries, the DC bus voltage, and the quality of energy delivered to the load. The PV system can easily track the MPP under uniform conditions using the P&O algorithm The DC bus voltage was maintained at 400 V by the storage device and the bidirectional converter despite all fluctuations in irradiance, as shown in Fig. 11a, and the RMS value remained constant (230 V), as shown in Fig. 11b.

We can see that the irradiance was at 700 W/m^2 before 0.4 s, the battery discharges since the power provided by the PV system is not enough for the load's demand and intervenes to supply the supplementary energy required by the load. The SOC of the battery due to these irradiance variations is shown in Fig. 11e, after 0.4 s irradiance becomes 1000 W/m^2, and the battery SOC becomes more stable. The load voltage is kept stable and closer to the sinusoidal form with a total harmonic distortion (THD) less than 1%, as illustrated in Fig. 11c, d.

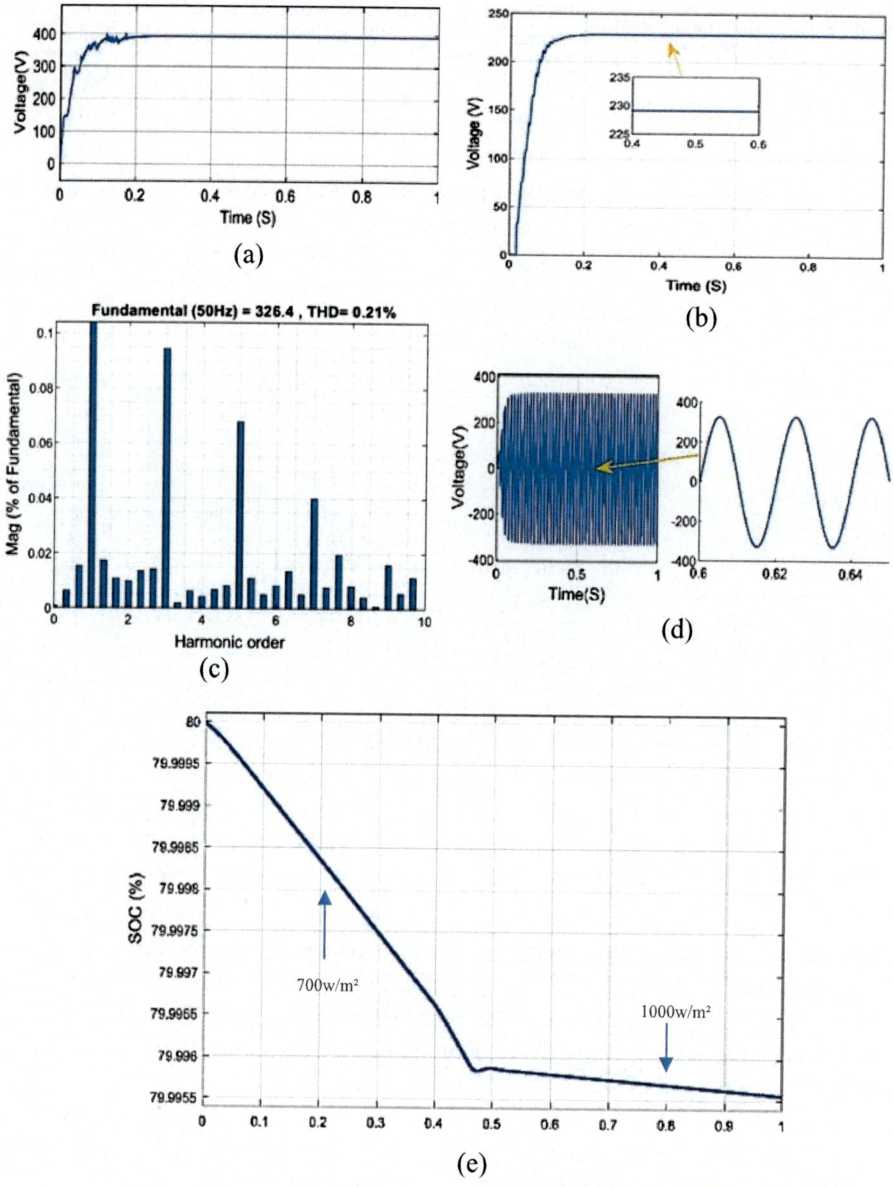

Fig. 11 **a** DC bus voltage, **b** RMS load voltage, **c** battery SOC level, **d** load voltage, **e** Harmonic spectrum of the load voltage

5.2 Extracted Power Under PS Conditions

The second scenario is to test the system under partial shading conditions with different patterns as illustrated in Table 2. These patterns are examined with FLC, which is compared with conventional P&O in order to evaluate the extracted power from PV panels. Figure 4a shows the P–V curve under different shading patterns.

Pattern 1 (Fig. 12b): in this case, the PV modules were irradiated with 300 W/m^2 at left corner, the global maximum power point (GMPP) is 1538 W/m^2, the shading loss is about 454 W, and the FLC was able to track the maximum power point, while the PO could not.

Pattern 2 (Fig. 12c): In this case, the system has been examined with bottom PV panels shaded with an irradiance of 600 W/m^2. The GMPP is 1292.5 W/m^2. The MPPT algorithms are not able to track the GMPP.

Pattern 3 (Fig. 12d): In this scenario, the system was tested using various PV array irradiance. The GMPP is 976.22 W/m^2. This pattern gives the lowest extracted power.

Uniform: the system has been tested under STC (1000 W/m^2), the GMPP is 1992 W/m^2. The results are presented in Fig. 12a.

The results in Fig. 12 shows that the GMPP cannot be tracked by the MPPT algorithms, and they stick at the first peak. Moreover, the proposed FLC has been

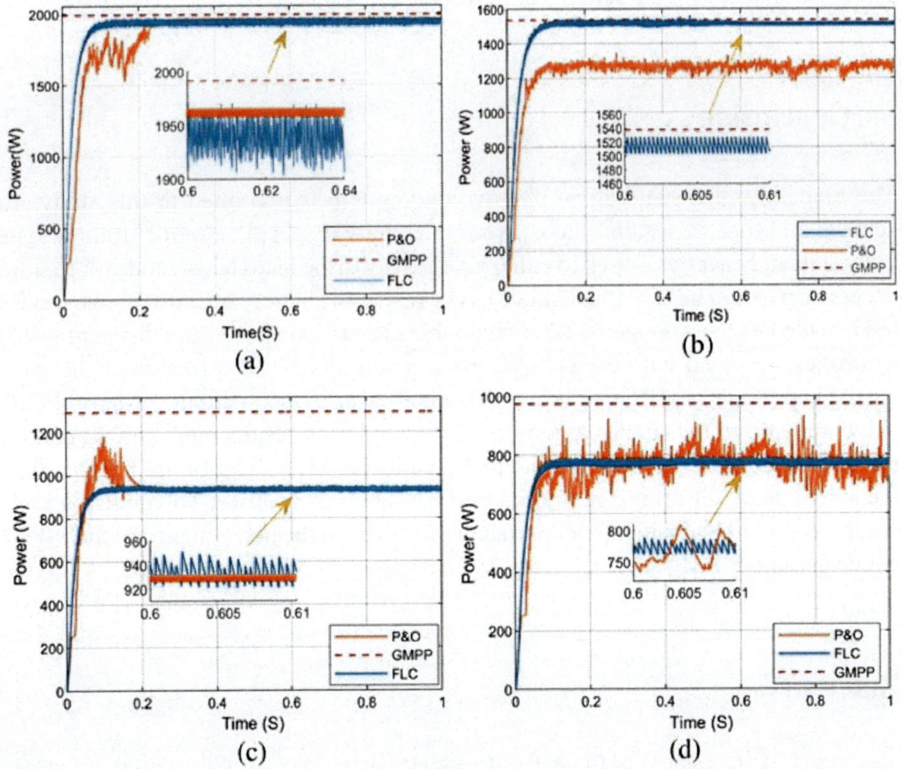

Fig. 12 **a** Extracted power for uniform irradiance, **b** extracted power for Pattern 1, **c** extracted power for pattern 2, **d** extracted power for pattern 3

Table 6 Parameters in various MPPT algorithms

		P_{MP}	Shading loss (w)	Fill factor
Pattern 1	P&O	1270	454	0.4859
	FLC	1515		0.5797
Pattern 2	P&O	930	699.5	0.3578
	FLC	934		0.3594
Pattern 3	P&O	773	1015.78	0.3007
	FLC	775		0.3015
Uniform	P&O	1962	–	0.7515
	FLC	1940		0.7431

compared with P&O for MPP, fill factor and shading loss. The Fill factor is defined as the ratio of maximum power to the product of the PV system's open-circuit voltage and short circuit current. If the fill factor value is close to one, the system is more efficient [6]. The shading losses represent the difference in power between the maximum power obtained from an array under STC and the total maximum power under PSC. These parameters for all patterns are presented in Table 6.

6 Conclusion

The modeling of a stand-alone photovoltaic system is presented in this study. To achieve an efficient system, the converters employ a specific control strategy. The bidirectional converter is used to either give energy to the load (boost mode) or to store excess energy in a battery (buck mode). As a result, the battery operates in two modes to keep the DC bus voltage stable. The double closed-loop proportional-integral (PI) controller was used with the DC-AC inverter to reduce THD. To extract the most available power from PV panels, the boost converter is used with the proposed FLC and compared to the (P&O) algorithm. The simulation results indicate that control techniques are robust. However, under PS conditions, MPPT algorithms fail to follow the global peak (GP) if it is not positioned first at the P–V curve. So, the fuzzy logic control must be used with other approaches to improve the performance of the MPPT controller under PSCs.

References

1. Ahmad, M.: Operation and Control of Renewable Energy Systems. Wiley (2018). https://doi.org/10.1002/9781119281733
2. Bakkar, M., Aboelhassan, A., Abdelgeliel, M., Galea, M.: PV Systems control using fuzzy logic controller employing dynamic safety margin under normal and partial shading conditions. Energies **14**(4), 841 (2021). https://doi.org/10.3390/en14040841

3. Nair, V.V., Ilango, K.: Microgrid control strategies for enhanced storage management. In: Proceedings of 2017 IEEE International Conference on Technological Advancements in Power and Energy: Exploring Energy Solutions for an Intelligent Power Grid, TAP Energy 2017, June 2018, pp. 1–5. https://doi.org/10.1109/TAPENERGY.2017.8397356
4. Javed, K., Ashfaq, H., Singh, R.: A new simple MPPT algorithm to track MPP under partial shading for solar photovoltaic systems. Int. J. Green Energy **17**(1), 48–61 (2020). https://doi.org/10.1080/15435075.2019.1686001
5. Chalh, A., Motahhir, S., El, A., Aboubakr, G., Hammoumi, E., Derouich, A.: Global MPPT of photovoltaic system based on scanning method under partial shading condition. SN Appl. Sci. (2020). https://doi.org/10.1007/s42452-020-2580-z
6. Verma, P., Garg, R., Mahajan, P.: Asymmetrical interval type-2 fuzzy logic control based MPPT tuning for PV system under partial shading condition. ISA Trans. **100**, 251–263 (2020). https://doi.org/10.1016/j.isatra.2020.01.009
7. Ngo, S., Chiu, C.S.: Simulation Implementation of MPPT Design under Partial Shading Effect of PV Panels (2020). https://doi.org/10.1109/ICSSE50014.2020.9219306
8. Thakran, S., Singh, J., Garg, R., Mahajan, P.: Implementation of PO algorithm for MPPT in SPV system. In: 2018 International Conference on Power Energy, Environment and Intelligent Control, PEEIC 2018, pp. 242–245 (2019). https://doi.org/10.1109/PEEIC.2018.8665588
9. Chaieb, H., Sakly, A.: Review and comparison of BAT and PSO MPPT's based algorithms for photovoltaic system. WSEAS Trans. Power Syst. **13**, 108–117 (2018)
10. Singh, O., Gupta, S.K.: A review on recent Mppt techniques for photovoltaic system. In: 2018 IEEMA Engineer Infinite Conference (eTechNxT), pp. 1–6 (2018)
11. Kermadi, M., Salam, Z., Ahmed, J., Berkouk, E.M.: An effective hybrid maximum power point tracker of photovoltaic arrays for complex partial shading conditions. IEEE Trans. Industr. Electron. **66**(9), 6990–7000 (2019). https://doi.org/10.1109/TIE.2018.2877202
12. Sankar, R., Velladurai, S., Rajarajan, R., Thulasi, J.A.: II. PV system description: maximum power extraction in PV system using fuzzy logic and dual MPPT control. In: 2017 International Conference on Energy, Communication, Data Analytics and Soft Computing, ICECDS 2017, June 2018, pp. 3764–3769. https://doi.org/10.1109/ICECDS.2017.8390168
13. Oufqir, F., Bendaoud, M., Chikh, K., Lokriti, A.: Modeling and control of a photovoltaic solar system using a storage and voltage stabilization battery for an efficient microgrid (2020). https://doi.org/10.1109/ICECOCS50124.2020.9314398
14. Hashim, N., Salam, Z., Johari, D., Nik Ismail, N.F.: DC-DC boost converter design for fast and accurate MPPT Algorithms in stand-alone photovoltaic system. Int. J. Power Electron. Drive Syst. (IJPEDS) **9**(3), 1038 (2018). https://doi.org/10.11591/ijpeds.v9.i3.pp1038-1050
15. da Rocha, M.V., Sampaio, L.P., da Silva, S.A.O.: Comparative analysis of MPPT algorithms based on Bat algorithm for PV systems under partial shading condition. Sustain Energy Technol Assess **40**, 100761 (2020). https://doi.org/10.1016/j.seta.2020.100761
16. Dhimish, M.: Assessing MPPT techniques on hot-spotted and partially shaded photovoltaic modules: comprehensive review based on experimental data. IEEE Trans. Electron Devices **66**(3), 1132–1144 (2019). https://doi.org/10.1109/TED.2019.2894009
17. Paul, A., Dey, B.K., Mandal, N., Bhattacharjee, A.: MATLAB/Simulink model of stand-alone Solar PV system with MPPT enabled optimized Power conditioning unit. In: 7th IEEE annual information technology, electronics and mobile communication conference, IEEE IEMCON 2016, no. 1, pp. 4–9 (2016). https://doi.org/10.1109/IEMCON.2016.7746319
18. Saleh, M., Esa, Y., Mhandi, Y., Brandauer, W., Mohamed, A.: Design and implementation of CCNY DC microgrid testbed. In: IEEE Industry Application Society, 52nd Annual Meeting: IAS 2016, pp. 1–7 (2016). https://doi.org/10.1109/IAS.2016.7731870
19. Motahhir, S., el Ghzizal, A., Sebti, S., Derouich, A.: Modeling of photovoltaic system with modified incremental conductance algorithm for fast changes of irradiance. Int. J. Photoenergy (2018). https://doi.org/10.1155/2018/3286479
20. Xiao, W.: Photovoltaic power. System (2017). https://doi.org/10.1002/9781119280408
21. Plotnikov, I., Atamankin, G.: Selecting components for bidirectional DC-DC converter of photovoltaic power supply system. In: Proceedings—2019 IEEE Russian Workshop on Power

Engineering and Automation of Metallurgy Industry: Research and Practice, PEAMI 2019, Oct. 2019, pp. 64–69. https://doi.org/10.1109/PEAMI.2019.8915133
22. Bhagiya, R.D., Patel, D.R.M.: PWM based double loop PI control of a bidirectional DC-DC converter in a standalone PV/Battery DC power system (2019). https://doi.org/10.1109/INDICON47234.2019.9028974
23. Dhaneria, A., Khambhadiya, H.: Design of AC side filter for grid tied solar inverter. In: 2019 4th IEEE International Conference on Recent Trends on Electronics, Information, Communication and Technology, RTEICT 2019—Proceedings, May 2019, pp. 1375–1378. https://doi.org/10.1109/RTEICT46194.2019.9016892
24. Istardi, D., Halim, B., Febriansyah, A.J.: High efficiency single phase inverter design. In: International Conference on Electrical Engineering, Computer Science and Informatics (EECSI), vol. 2017. https://doi.org/10.1109/EECSI.2017.8239137
25. Mishra, J., Das, S., Kumar, D., Pattnaik, M.: Performance comparison of PO and INC MPPT algorithm for a stand-alone PV system. In: 2019 Innovations in Power and Advanced Computing Technologies, i-PACT 2019, no. 1, pp. 1–5 (2019). https://doi.org/10.1109/i-PACT44901.2019.8960005
26. Yuwanda, R.I., Prasetyono, E., Eviningsih, R.P.: Constant power generation using modified MPPT PO to overcome overvoltage on solar power plants. In: Proceedings—2020 International Seminar on Intelligent Technology and Its Application: Humanification of Reliable Intelligent Systems, ISITIA 2020, Jul. 2020, pp. 392–397. https://doi.org/10.1109/ISITIA49792.2020.9163685
27. Eltamaly, A.M., Abdelaziz, A.Y.: Modern Maximum Power Point Tracking Techniques for Photovoltaic Energy Systems (2012) [Online]. Available: http://www.eurekaselect.com/102081/volume/1
28. Ali, M.N., Mahmoud, K., Lehtonen, M., Darwish, M.M.F.: Promising mppt methods combining metaheuristic, fuzzy-logic and ANN techniques for grid-connected photovoltaic. Sensors (Switzerland) **21**(4), 1–18 (2021). https://doi.org/10.3390/s21041244

Comparison of the Efficiency of ANN Training Algorithms for Tracking the Maximum Power Point of Photovoltaic Field

Ncir Noamane, Sebbane Saliha, and Nabil El Akchioui

Abstract Global energy demand is growing very rapidly. Suddenly, due to industrial development, the supply of fossil natural resources is shrinking, which poses a key issue for the future of the world. For this reason, the development of renewable energies remains an economic and environmental solution to achieve the expected objectives. However, due to changes in weather conditions, energy efficiency is one of the biggest issues in renewable energy development. In this article, we put a Photovoltaic (PV) panel under Standard Test Conditions (STC) to measure and improve its performance in terms of speed, stability, and precision based on Artificial Intelligence (AI) and more precisely Artificial Neural Networks (ANN). In this case, MATLAB software was used to simulate several training algorithms, and the results were compared in terms of performance to track the maximum power point (MPPT) of the photovoltaic panel.

Keywords Artificial intelligence · Artificial neural network · Gradient descent · Bayesian regularization · Levenberg–Marquardt · Scaled conjugate gradient · Photovoltaic · Maximum power point tracking

1 Introduction

Photovoltaic solar energy is considered one of the most demanded energy sources in the world, thanks to its eco-efficiency and the fact that it solves the economic and environmental problems associated with the use of others such as natural gas, oil, and fossil fuels [1].

N. Noamane (✉) · S. Saliha · N. El Akchioui
Faculty of Science and Technology, University Abdelmalek Essaadi/Al Hoceima, Tetouan, Morocco
e-mail: noamane.ncir@etu.uae.ac.ma

S. Saliha
e-mail: saliha.sebbane@etu.uae.ac.ma

N. El Akchioui
e-mail: n.elakchioui@uae.ac.ma

M. Bendaoud et al. (eds.), *The Proceedings of the International Conference on Electrical Systems & Automation*, https://doi.org/10.1007/978-981-19-0035-8_2

On the other hand, compared to fossil fuels, the energy efficiency of photovoltaic solar panels is too low on the pretext that there are certain climatic faults and behaviors [2], which affect the performance of photovoltaic panels. Therefore, the photovoltaic current or voltage is always unstable, that is, the efficiency of the photovoltaic panel is always low. In this fact, the integrated maximum power point tracking (MPPT) system is an effective way to increase the gain of power generation. In the literature, a variety of algorithms are used to solve this problem, the first is the perturbation and observation (P&O) method, Incremental conductance (INC), Fractional Short Circuit Current Algorithm (FSCC), or Fractional Open Circuit Voltage Algorithm (FOSC). Therefore, a more sophisticated method is needed to extract the maximum performance from photovoltaic panels, of which we have chosen to integrate the theory of artificial intelligence and more precisely the Artificial Neural Networks (ANN), which have been discussed by several researchers [3–5], to solve the complications of classical algorithms.

2 Modeling of Photovoltaic Panels

The electrical modeling of PV modules is a very important step, the goal of which is to increase energy efficiency during energy production. For this reason, several PV models strongly influence the operation of PV modules after their manufacture [6].

Therefore, the main objective when modeling PV is to have an output power (P_{PV}) which guarantees the energy requirement of a load of a PV installation. In addition, the two parameters that allow PV modules to operate are temperature (T) and irradiance (G), so increasing irradiation and decreasing temperature implies increasing the current of the photon (I_{ph}). Figure 1 shows the diagram of an equivalent circuit of a PV model.

Indeed, PV technologies are based on the photoelectric effect or in other words the PV effect, which presents the idea that PV cells and modules are naturally autonomous energy generators, which require no fuel and which practically rely on an infinity of solar energy [7].

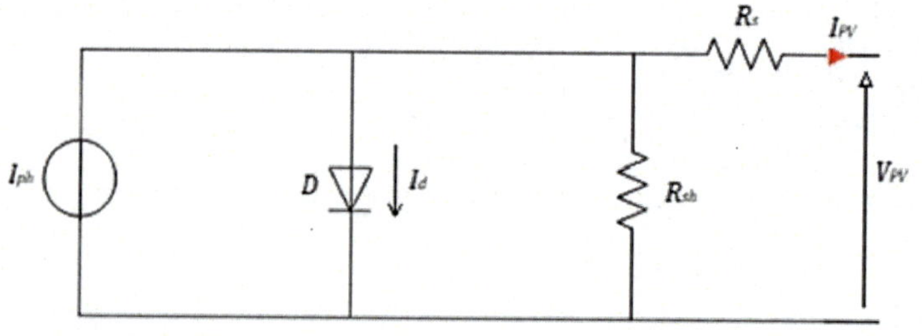

Fig. 1 Equivalent circuit of a PV model

$$I_{PV} = I_{ph} - I_0\left[e^{\frac{V_{PV}+R_S.I_{PV}}{V_t.a}} - 1\right] - \frac{V_{PV} + R_s.I_{PV}}{R_{sh}} \tag{1}$$

$$V_t = \frac{N_s.K.T}{q} \tag{2}$$

3 Maximum Power Point Tracker (MPPT)

The MPPT is a system that controls the current and the voltage at the output of the photovoltaic panels, which offers a very high performance, thus it calculates in real time the optimal power point to deliver all the power to the batteries or network thanks to their microprocessor and load algorithms. There are several MPPT algorithms that can be separated into different groups such as classical methods, distributed approaches, methods based on evolutionary algorithms as well as methods based on artificial intelligence [8].

4 Artificial Intelligence (AI)

Artificial intelligence is any algorithmic computer technology whose goal is to solve complex problems by drawing inspiration from human resonance and by simulating human capacities such as perception and reasoning without accessing the notion of consciousness. The two disciplines of artificial intelligence are presented as follows in Fig. 2.

Fig. 2 The disciplines of AI

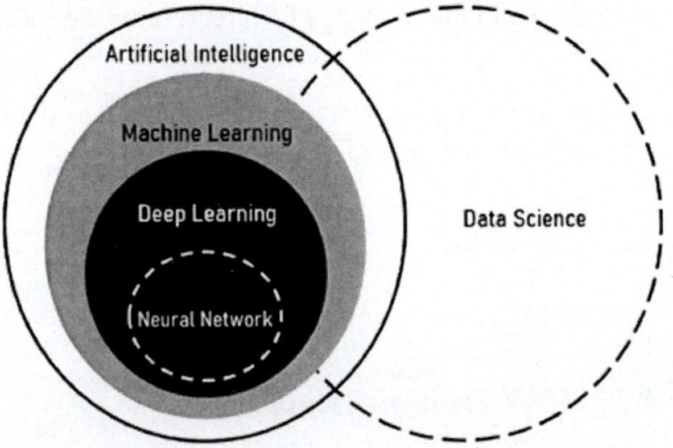

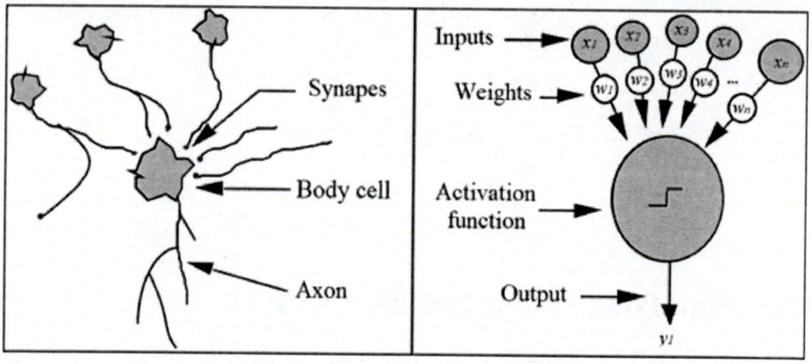

Fig. 3 Representation of biological neuron and artificial neuron

4.1 Artificial Neural Networks (ANN)

Deep learning (DL) is one of the machine learning (ML) subfields of study that is based on the construction of artificial neural networks (ANN). It applies to systems that are more complex and have much larger amounts of data than ML.

The idea of creating ANNs mainly based on reconstructing concepts of the human brain, which contains thousands or millions of biological neurons, thus learning these concepts and simulating their functioning [9–11].

Otherwise, a single biological neuron receives signals transmitted by other neurons (this is the dendritic-synaptic interaction). The representation of this artificial neuron is illustrated in Fig. 3.

In this case, we can translate biological neurons in a simple way through artificial representations, including.

- The inputs are represented as a vector representing the dendrites,
- Weights that take on the role of synapses,
- The activation function, which represents the cell body,
- The output, which represents the axon.

The artificial representation of a neuron can be converted in the form of a mathematical equation (3):

$$\hat{y} = f(\langle \omega, \rangle + b) \tag{3}$$

4.2 ANN Training Algorithms

Gradient descent (GD). Gradient descent is an algorithm that converges to the minimum of the used cost function, which is normally a convex function facing the valley and is given by (4):

$$J(a) = \frac{1}{2m} \sum_{i=1}^{m} \left(f\left(x^{(i)} - y^{(i)}\right)^2 \right) \tag{4}$$

The algorithm selects randomly the problem's parameters from the search system. For example, in the case of two parameters a and b, we use alpha as the step size to calculate the partial derivative of J(a) with respect to each parameter until we find the small value of the cost function. Taking into account the number of parameters of the studied system. Formula (5) describes mathematically all parameters used to reach the optimum based on gradient descent.

$$a_{i+1} = a_i - \alpha \frac{\partial J a_i}{\partial a} \tag{5}$$

Levenberg–Marquardt (LM). This type of training algorithm combines the idea of descent gradient and the Gauss–Newton method for adapting the weights of the ANNs using the mathematical model (6–7) [12].

$$p^{k+1} = p^k + \left(X^{Tk}.W.X^k\right)^{-1}.X^{Tk}.W.\left(y - i^{-k}\right) \tag{6}$$

It was originally designed for nonlinear parameter estimation problems, but it has also been proven to be used to solve linear problems under harsh conditions [13].

$$p^{k+1} = p^k + \left(X^{Tk}.W.X^k + \mu^k \Omega^k\right)^{-1}.X^{Tk}.W.\left(y - i^{-k}\right) \tag{7}$$

Bayesian Regularization (BR). BR is used as a training algorithm for ANN by adding to the backpropagation-learning algorithm some regularization techniques, in order to obtain a small error value and to have a training algorithm that converges faster than the backpropagation algorithm [14].

ANN training with BR incorporates Bayes theorem [15]. The BR algorithm requires the Hessian matrix of the objective function [16]. The objective function is given by the Mean Squared Error (MSE) cost function and the regularization by the sum of the weights squared (8) and (9):

$$F = \alpha E_W + \beta E_D \tag{8}$$

$$E_W = \sum_{k=1}^{l} \sum_{i,j=1}^{m} \left(w_{ijk}\right)^2 \tag{9}$$

where E_W is the squared sum of the weights, E_D is the MSE cost function on the input and target data and finally α and β are the parameters of the objective function.

The weights of the ANN that follow a probability distribution are given in (10):

$$f(W|D,\alpha,\beta,M)\frac{\exp^{(-(\alpha E_W+\beta E_D))}}{Z_W(\alpha)Z_D(\beta)}\frac{\exp^{(-F(w))}}{Z_F(\alpha,\beta)} \tag{10}$$

where M is the used ANN model where: $Z_F(\alpha,\beta)=\int\exp^{(-F)}dw$; $Z_D(\beta)=2\pi/\beta$ $\beta=1/\sigma^2$; $Z_W(\alpha)=\int\exp^{(-\alpha E_W)}dw$.

Scaled Conjugate Gradient (SCG). This algorithm is a derivation of the Conjugate Gradient Learning (CG) algorithm class [17], to train ANNs. Training with the CG provides more robust convergence than the GD [18]. The SCG uses the same CG strategy, but the search direction and step size are chosen efficiently using information from the second-order approximation [19], represented by (11):

$$E(w+y)\approx E'(w)+E'(w)^T+\frac{1}{2}y^TE''(w)y \tag{11}$$

where $E(w)$ is a global function depending on all the weights and biases attached to the neural network, $E'(w)$ represent the gradient and $E''(w)$ represent the Hessian matrix. The quadratic approximation of E in a neighborhood of a point w is given by Eq. (12):

$$E_{qw}(y)=E'(w)+E'(w)^T+\frac{1}{2}y^TE''(w)y \tag{12}$$

To determine the minimums of E_{qw}, we use Eq. (13):

$$E'_{qw}(y)=E''(w)y+E'(w)=0 \tag{13}$$

5 Configuration of the Proposed Approach

In order to evaluate each training algorithm cited above in the previous section and to summarize the efficiency of models based mainly on those algorithms and the proposed ANN architecture, a procedure is followed to achieve the objective of this work following six steps. The procedure is starting by defining the main problem, collecting data, analyzing problem and data, validating and evaluating the created model, and ending by testing the model under several conditions to prove his capability to predict the best results. In fact, the proposed approach is tested to predict the best duty cycle in the purpose of reaching the MPPT of the chosen PV panel.

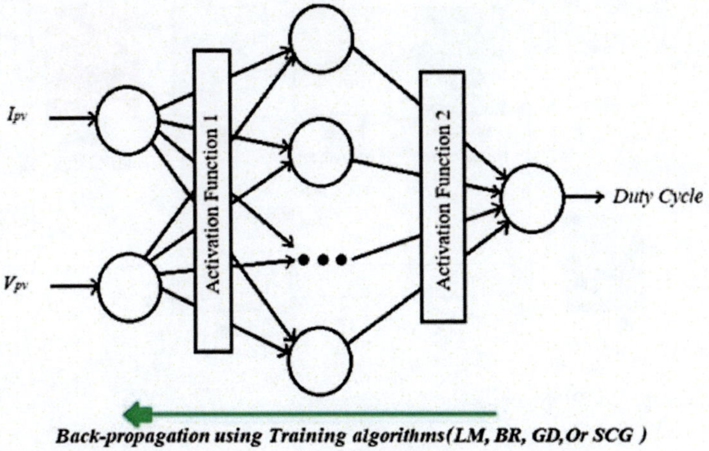

Fig. 4 Proposed ANN Architecture

5.1 Proposed ANN Architecture

In this Article, the created ANN Model is based on MLP architecture of the feed-forward ANN of three layers of nodes, including input layer with 2 features representing the current and the voltage of the used PV panel, hidden layer, and output layer with 1 feature representing duty cycle as shown in Fig. 4. In fact, all layers are fully connected continuously using weights, with no backward connection and non-adjacent layers. In addition, all layers are using non-linear functions to convert data into a mathematical form. Furthermore, the model is trained using supervised learning with a technique called backpropagation that allows the model to change the values of weights at each iteration to evaluate the variation and reach the minimum of the objective function, generally the Mean Square Error (MSE). As well, this supervised learning is based on algorithms cited in the previous section.

5.2 Proposed PV System

The proposed system is tested under Standard Test condition (STC) (T = 25°, Ir = 1000 W/m^2) and it consists of the aforementioned PV panel, boost converter, MPPT system, and simple load to evaluate the performance of each algorithm. In addition, the different results are simulated by MATLAB-Simulink software as shown in Fig. 5.

Dataset is collected from a simulation of the maximum performance of the used PV module (1STH-215-P) with an architecture consisting of a series of 5 modules of the same type linked in parallel and it contains 36,301 examples for training the ANN model.

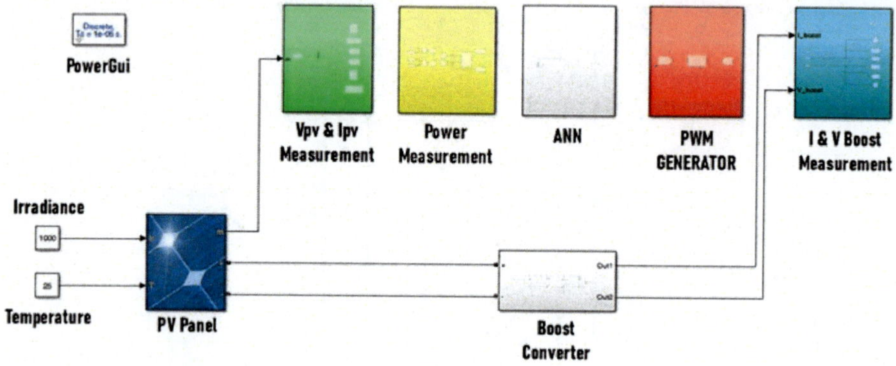

Fig. 5 The proposed PV system

6 Results and Discussion

6.1 *Simulation Results of the Proposed ANN Model*

In this paper, a comparison between four algorithms (GD, LM, BR and SCG) is discussed to choose the best configuration for the ANN model. Parameters such as regression, MSE, Gradient, Momentum parameters, and validation are often used to determine the performance accuracy of any algorithm for a dataset. In addition to that, the regression represents the efficiency of the prediction and it is evaluated by the error that is calculated at the output of the target. To train the ANN model, we choose to separate the dataset set into 3 Phases: 70% Training, 15% Validation, and 15% test.

The performance of each algorithm is evaluated based on the number of neurons, training algorithms, and the MSE fitness function. The results of this simulation are presented in Table 1 (Fig. 6).

After training the ANN model with the different algorithms and according to several selection criteria of learning parameters, we can notice that the algorithm (BR) has a very high precision (99.999%) using 30 neurons in a hidden layer. However, we can notice on the basis of the comparison between results obtained by BR with the other algorithms, that LM and SCG algorithms present some good results as (99.996%) for the LM that shows a faster convergence towards the optimum using just a small number of neurons as indicated, and (99.840%) for the SCG. In contrast, the GD algorithm shows its inability to converge to the optimum solution. Figure 6 represents more clearly the variation of the MSE fitness curve of the best configuration obtained whether in training, validation, or test phases. In fact, the best validation performance has reached 2.0254 e−06 in 500 iterations as shown in below.

Table 1 Simulation results of each training algorithm

Training algorithm	Neurons number	Iterations number	MSE	Accuracy (%)
Gradient descent (GD)	5	10	104.9085	−1.030
	10	10	146.7447	−1.450
	15	10	2.2893 e+03	−22.880
	20	10	125.2027	−1.240
	25	10	585.4966	−5.840
	30	10	878.9677	−8.780
Bayesian regularization (BR)	5	500	4.5671 e−05	99.995
	10	500	3.3839 e−05	99.996
	15	500	2.0920 e−04	99.979
	20	500	3.7644 e−05	99.996
	25	500	1.7124 e−05	99.998
	30	500	2.6545 e−06	99.999
Levenberg–Marquardt (LM)	5	500	5.9904 e−04	99.940
	10	500	9.5012 e−05	99.990
	15	500	3.1084 e−05	99.996
	20	500	3.1428 e−05	99.996
	25	500	3.6324 e−05	99.996
	30	500	7.7499 e−05	99.992
Scaled conjugate gradient (SCG)	5	374	0.0076	99.240
	10	362	0.0049	99.510
	15	500	0.0021	99.790
	20	364	0.0054	99.460
	25	500	0.0016	99.840
	30	479	0.0077	99.230

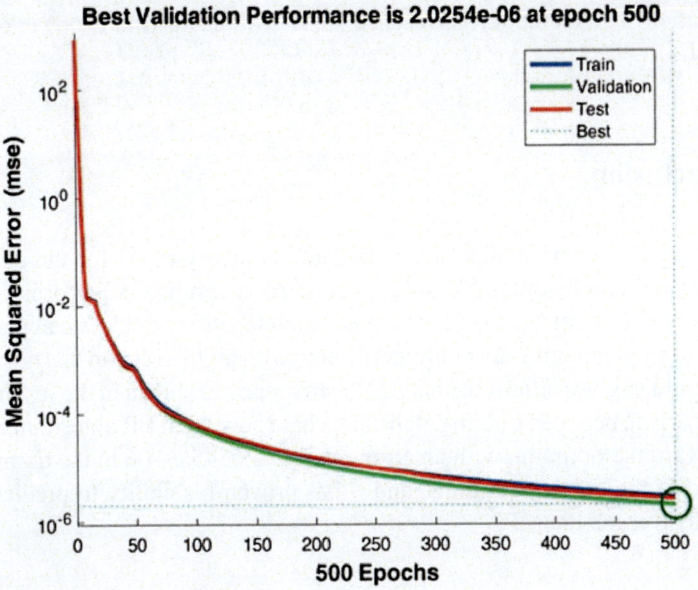

Fig. 6 Best Performance for the BR Algorithm (30 neurons)

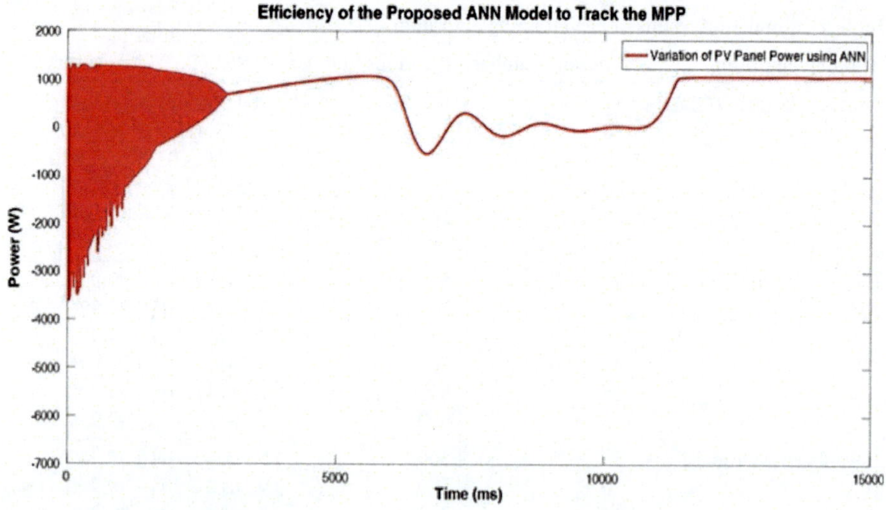

Fig. 7 Efficiency of the proposed ANN model to track the MPP

6.2 Efficiency of the ANN Model to Track the MPP

For any MPPT algorithm, tracking efficiency and tracking time are very important to get the most benefit from the PV system. Tracking efficiency is the ratio between the maximum power tracked by the algorithm and the maximum power of the PV configuration. From the results represented in Fig. 7, it is shown that the best obtained configuration of ANN model has reached 99.97% of the maximum power of the used PV panel.

7 Conclusion

In this research, we have chosen to integrate the theory of AI for the objective of increasing the efficiency of PV solar panels. As a result, the performance of the proposed ANN model shows that the major complications of classic algorithms are resolved by implementing an intelligent system with high precision that decreases the oscillations and gives almost the maximum efficiency as shown in the results. In fact, this comparison between training algorithms has shown that BR algorithm converges better toward the optimum with an error equal to 2.6545 e−06 in the training phase compared to the other algorithms, and it has proven his ability to predict the best duty cycle to reach the MPP.

References

1. Yap, K.Y., Sarimuthu, C.R., Lim, J.M.Y.: Artificial intelligence based MPPT techniques for solar power system: a review. J. Mod. Power Syst. Clean Energy. **8**(6), 1043–1059 (2020)
2. Anzalchi, A., Sarwat, A.: Artificial neural network based duty cycle estimation for maximum power point tracking in photovoltaic systems. In: SoutheastCon 2015, Fort Lauderdale, FL, USA, pp. 1–5, (2015)
3. Braspenning, P.J., Thujisman, F., Weijters, A.J.M.M.: Artificial neural networks an introduction to ANN theory and practice, 1re éd. Springer, Berlin (1995)
4. Dhekekar, R.S., Srikanth, N.V.: ANN Controlled VSC STATCOM with Harmonic Reduction for VAR Compensation, vol. 2, no 1, p. 9 (2012)
5. Yaichi, M., Fellah, M.-K., Mammeri, A.: A neural network based MPPT technique controller for photovoltaic pumping system. Int. J. Power Electron. Drive Syst. IJPEDS. **4**(2), 241–255 (2014)
6. Sredenšek, K., Štumberger, B., Hadžiselimović, M., Seme, S., Deželak, K.: Experimental validation of a thermo-electric model of the photovoltaic module under outdoor conditions. Appl. Sci. **11**(11), 5287 (2021)
7. Sedaghati, F., Nahavandi, A., Badamchizadeh, M.A., Ghaemi, S., Abedinpour Fallah, M.:PV maximum power-point tracking by using artificial neural network. Math. Probl. Eng. **2012**, 1–10 (2012)
8. Motahhir, S., El Hammoumi, A., El Ghzizal, A.: The most used MPPT algorithms: Review and the suitable low-cost embedded board for each algorithm. J. Clean. Prod. **246**, 118983 (2020)
9. Pan, H., Niu, X., Li, R., Dou, Y., Jiang, H.: Annealed gradient descent for deep learning. Neurocomputing **380**, 201–211 (2020)
10. Alonso-Montesinos, J.: The use of ANN and conventional solar-plant meteorological variables to estimate atmospheric horizontal extinction. G Lo 13 (2021)
11. Lecun, Y.: Gradient-based learning applied to document recognition. Proc. IEEE **86**(11), 47 (1998)
12. Modest, M.F.: Inverse radiative heat transfer. In: Radiative Heat Transfer, pp. 729–742, Elsevier (2003)
13. Puig-Arnavat, M., Bruno, J.C.: Artificial neural networks for thermochemical conversion of biomass. In: Recent Advances in Thermo-Chemical Conversion of Biomass, pp. 133–156. Elsevier (2015)
14. Kayri, M.: Predictive abilities of Bayesian regularization and Levenberg–Marquardt algorithms in artificial neural networks: a comparative empirical study on social data. Math. Comput. Appl. **21**(2), 20 (2016)
15. Burden, F., Winkler, D.: Bayesian regularization of neural networks. In: Livingstone, D.J. (éd.) Artificial Neural Networks, vol. 458, pp. 23–42. Humana Press, Totowa, NJ (2008)
16. Sariev, E., Germano, G.: Bayesian regularized artificial neural networks for the estimation of the probability of default. Quant. Finance **20**(2), 311–328 (2020)
17. Nayak, S.: Scaled conjugate gradient backpropagation algorithm for selection of industrial robots. Int. J. Comput. Appl. **6**(7) (2017)
18. Forouzanfar, M., Dajani, H.R., Groza, V.Z., Bolic, M., Rajan, S.: Comparison of feed-forward neural network training algorithms for oscillometric blood pressure estimation. In: 4th International Workshop on Soft Computing Applications, Arad, Romania, pp. 119–123 (2010)
19. Møller, M.F.: A scaled conjugate gradient algorithm for fast supervised learning. Neural Netw. **6**(4), 525–533 (1993)

References

Development of an MPPT Control Based on Fuzzy Logic for a Photovoltaic System

S. Mountassir, S. Sarih, and M. Tajer

Abstract Today, the connection of electricity production systems from renewable sources is more than ever a necessity as the technologies used in this type of conversion are more and more accessible and at more competitive prices. These techniques have been optimized over the years either by optimizing photovoltaic cells, heat transfer liquids or asynchronous generators for example, or once again, by optimizing tracking control algorithms or power electronic components integrated into inverters. In this paper, the FLC strategy is directly integrated in the MPPT to control the DC-DC boost converter and MATLAB Simulink tool is intended for the simulation. The FLC demonstrates good performance and better waveform and response in varying environmental conditions.

Keywords P&O · Fuzzy logic · Boost converter · FLC·MPPT · PV system · Matlab/simulink

1 Introduction

The techniques and strategies of control of the stages of inverters to optimize the output waveform have become more and more widespread in the world and especially, given the new directives of the codes of electrical networks which put conditions for access to high, medium, and low voltage electrical energy distribution networks.

These conditions limit and regulate the technical parameters of the various electrical quantities to guarantee an injection of the energy produced in the best conditions of use and without disturbing the quality offered by the distributors to the end customers and thus to prevent momentary interruptions of the supply, voltage dips control, harmonics in the network and flickers.

S. Mountassir (✉) · S. Sarih · M. Tajer
Systems and Applications Engineering Laboratory (LISA), National School of Applied Sciences of Marrakech, Cadi Ayyad University, Marrakech, Morocco

M. Tajer
e-mail: a.tajer@uca.ac.ma

M. Bendaoud et al. (eds.), *The Proceedings of the International Conference on Electrical Systems & Automation*, https://doi.org/10.1007/978-981-19-0035-8_3

The contribution of this paper lies in the direct integration of an improved fuzzy logic command into the MPPT process [1, 2]. All simulations are performed using MATLAB/SIMULINK software.

In the present document, we propose a developed Fuzzy logic-based control to achieve the maximum power point tracking. The paper is structured as follows: the first part explains a simplified model of the PV system with a DC-DC boost converter, the second part describes the FLC strategy, and gives an analysis of the simulation results.

2 Block Diagram of a Photovoltaic System

The block diagram of the MPPT control of a system for producing photovoltaic electricity is composed of the photovoltaic panel module, the DC-DC boost converter, the MPPT controller, and the load to be powered. The output voltage of the Photovoltaic panel supplies the DC boost converter according to which a direct load is connected. Measurements of panel voltage and/or current are introduced to the input MPPT controller, then the MPPT algorithm [3]. The following figure illustrates the system's block diagram (Fig. 1).

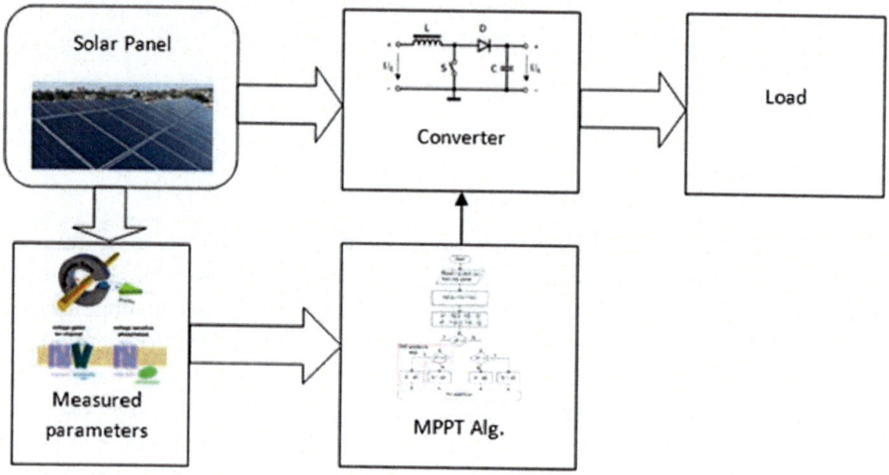

Fig. 1 Block scheme of a photovoltaic system

3 Photovoltaic Panel Model

Considered photovoltaic block consists of an array of photovoltaic (PV) modules. The matrix is made up of parallel strings of modules, each string comprising modules connected in series.

The Photovoltaic system object of our simulation is a five-parameter model using an I_L current source (current generated by solar irradiation), a diode, a series resistor R_s and a shunt resistor R_{sh} to characterize the dependent features of the module illumination and temperature [3] (Fig. 2).

The equivalent diagram is as follows:

The equation of the solar PV is given as below:

$$I = I_L - I_{R_{sh}} - I_d \tag{1}$$

$$I = I_L - I_{R_{sh}} - I_s e^{\left(\frac{q(V+R_S I)}{akT}\right)-1} \tag{2}$$

$$I = I_L - \left(\frac{V + R_s I}{R_{sh}}\right) - I_s e^{\left(\frac{q(V+R_S I)}{akT}\right)-1} \tag{3}$$

where

a diode ideality factor
T temperature
Is reverse saturation current
q electric charge.

The simulation subject of this paper is based on an ascending adjustment of the irradiation from 600 to 1000 W/m^2, and a fixed temperature of 25 °C, for an MPPT based on the fuzzy logic algorithm.

Fig. 2 A photovoltaic panel's circuit model

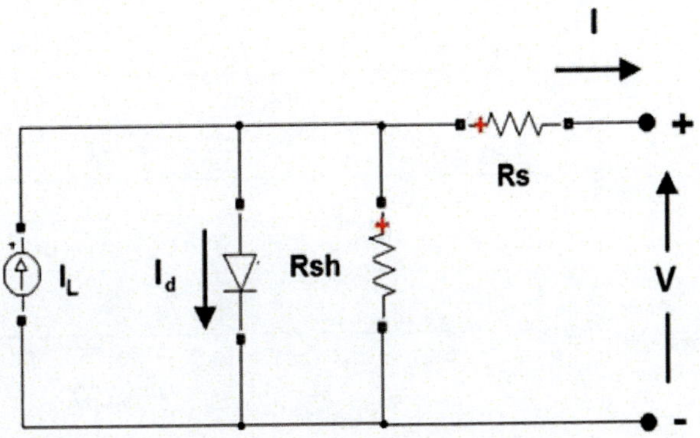

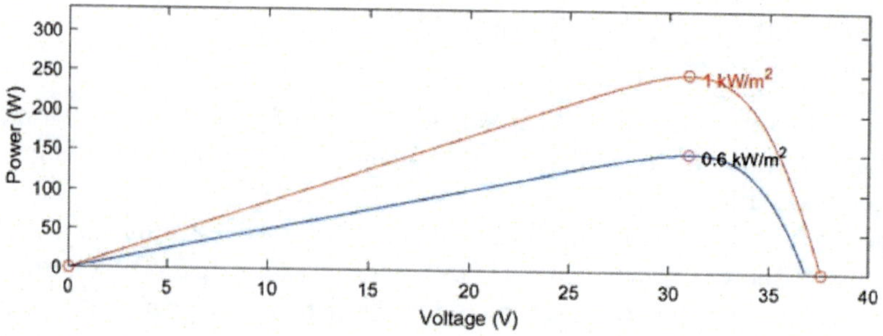

Fig. 3 Power profile as a function of voltage for variable solar irradiation

In a second step, a downward adjustment of the irradiation from 1000 to 600 W/m^2, and a fixed temperature of 25 °C, was carried out for the MPPT based also on the fuzzy logic algorithm.

The opposite case was also tested for the same MPPT control blocks, namely, continual solar irradiation at 1000 W/m^2 and the stepwise variation of the temperature from 25 to 50 °C ascending and likewise descending.

The fuzzy logic loop diagram was optimized for this application.

The following graph depicts the power profile as a function of voltage (Fig. 3) [4].

The following graph represents the current's profile as a function of voltage (Fig. 4).

The characteristic curves of the case of fixed solar exposure and a temperature variation of the PV are given below and the variations are given in steps to allow characterization of the dynamic behavior of the system.

The characteristic of the power according to the voltage is typically given by (Fig. 5).

The following graph (Fig. 6) represents the current's characteristic as a function of voltage.

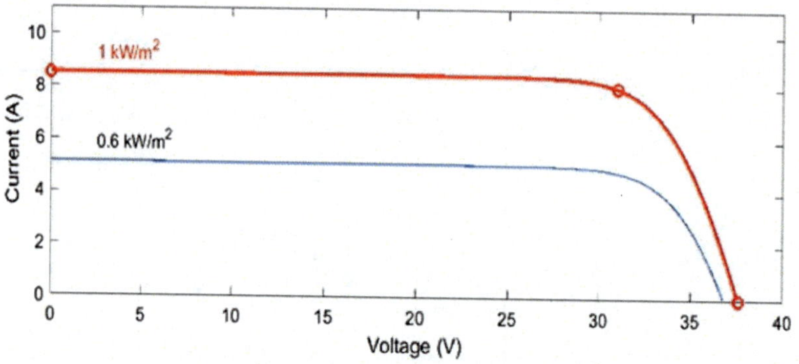

Fig. 4 Current as a function of voltage profile for variable solar irradiation

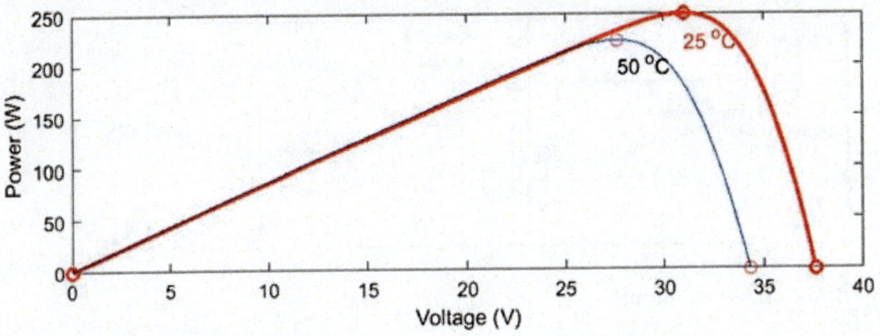

Fig. 5 Power characteristic as a function of voltage for a variable temperature

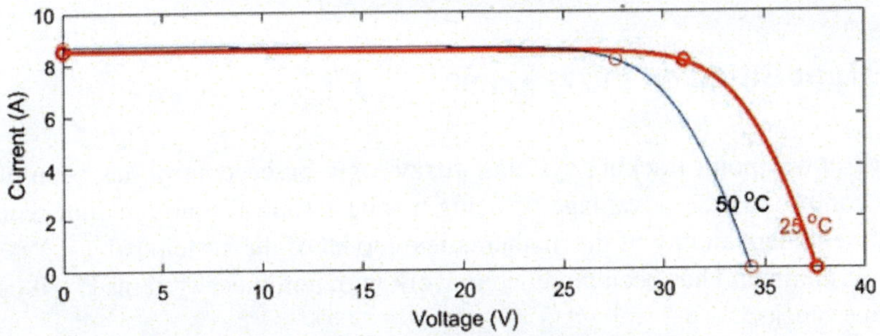

Fig. 6 Current as a function of voltage profile for a variable temperature

The characteristic curves are typical and make it possible to optimize the various strategies used and to refine and tune the settings [5].

4 Boost Converter

The boost converter circuit employed in this simulation is a simple DC-DC power converter, with a high-frequency switch and a diode, operating in an asynchronous mode which raises the voltage from its input to the output.

The circuit diagram is represented in the illustration in Fig. 7.

Where "V_{in}" is the voltage of the PV panel, and the command signal is generated by the MPPT module.

The transfer function of the boost converter in continuous conduction mode configuration is:

$$\frac{V_{out}}{V_{in}} = \frac{1}{1-\alpha} \tag{4}$$

where α is the duty cycle.

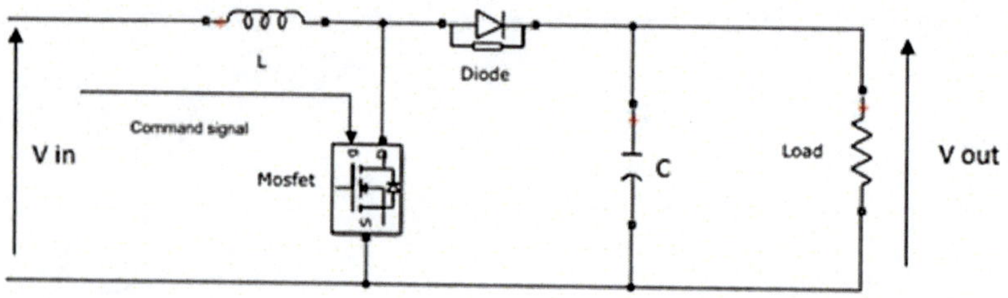

Fig. 7 Boost converter circuit

The command signals drive the gate of the power switch (MOSFET).

5 Mppt Based on Fuzzy Logic

In overpower point tracking systems, fuzzy logic-based control has been utilized. This control has the advantage of being a robust control that does not require an accurate understanding of the mathematical model of the system [6].

This command has been modified to work with nonlinear systems [7]. It operates in three blocks:

1. Fuzzification,
2. Inference
3. Defuzzification.

Where the fuzzification block permits the conversion of physical input variables into fuzzy sets. [8–10].

Figure 8 the diagram of the fuzzy logic used in this paper

In this paper, we have considered two inputs:

$$\text{Input 1:} = \frac{dP}{dV}$$

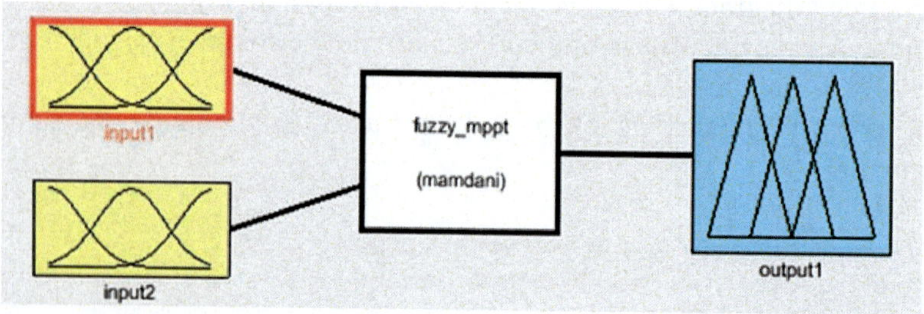

Fig. 8 Diagram of the fuzzy logic used

$$\text{Input 2} := \frac{de}{dt}$$

They are defined as follows:

$$e = \frac{dP}{dV} = \frac{P(k) - P(k-1)}{V(k) - V(k-1)} \quad (5)$$

$$de = e(k) - e(k-1) \quad (6)$$

The fuzzy sets are as follows:

N negative,
Z zero,
P positive.

For the Defuzzification, we have considered the centroid method over the Mamdani inference [11].

In our simulation, we have considered for the inputs and outputs, triangular membership functions so as not to make the equivalent system more complex and to decrease the running time (Fig. 9).

In the next step, while defining the membership rules, we construct logical relationships between the inputs and outputs. Then we make a table of rules (Table 1).

The rules foundation is shown in the table below (9 rules) set for the fuzzy logic mechanism considered:

6 Simulation Results and Discussion

The simulation was performed based on the FLC mentioned above, with a temperature set point fixed at 25 °C, and a solar irradiation set point in steps of 600–1000 W/m^2, and the opposite case by fixing the temperature at 25 °C and applying solar irradiation set point of 1000–600 W/m^2 to verify the system's stability in various scenarios and be able to compare behavior in various setups [4] (Figs. 10, 11, 12 and 13).

The general parameters are given in Table 2.

The simulation results are given by the curves below.

- First case

The first case treated is an unchanging temperature of 25 °C and an irradiation incremented from 600 to 1000 W/m^2.

The above illustration represents the response of the MPPT based on the FLC algorithm, the response based on the algorithm using fuzzy logic is given in red. The

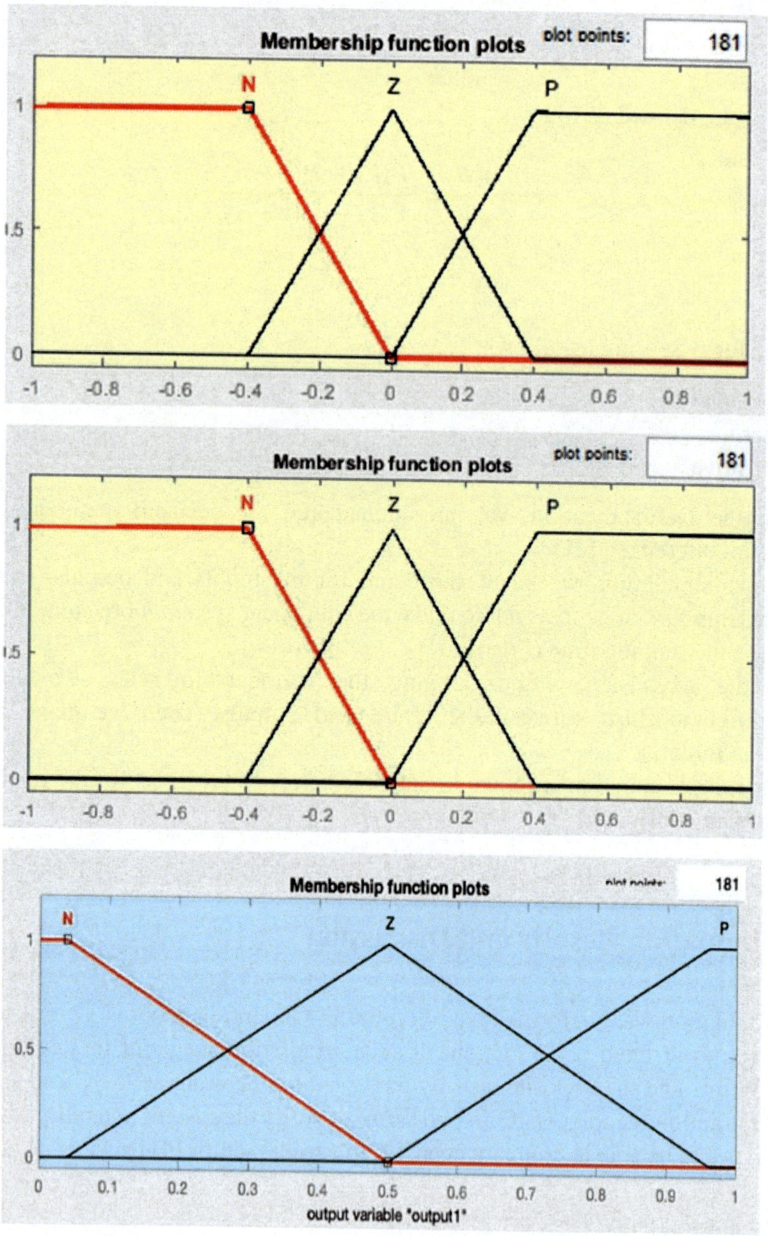

Fig. 9 **a** Membership functions plots for the variable: *e*, **b** membership functions plots for the variable: d*e*, **c** membership functions plots for the variable α

Table 1 Table of the output of the fuzzy controller α

		de		
		N	Z	P
e	N	N	N	Z
	Z	N	Z	P
	P	Z	P	P

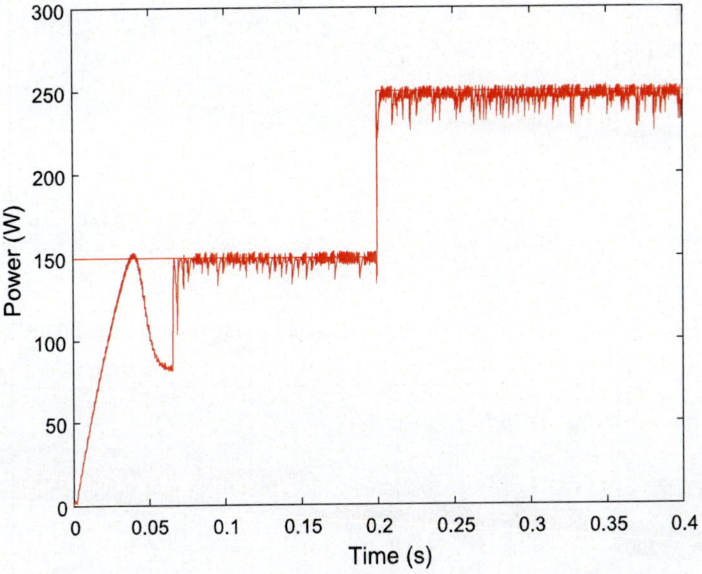

Fig. 10 The evolution of output powers for irradiation varying from 600 to 1000 W/m^2

behavior using fuzzy logic has fewer ripples and allows better response speed to the system than other strategies.

- Second case

The second case treated is a constant temperature of 25 °C and solar irradiation decremented from 1000 to 600 W/m^2:

- Third case

The third case treated is a constant solar irradiation of 1000 W/m^2, and a temperature decreasing from 50 to 25 °C.

Figure 12 represents the response based on the algorithm using fuzzy logic. The behavior using fuzzy logic has fewer ripples and allows quicker response to the system. It also demonstrates the effect of increasing temperature parameters regarding the power produced.

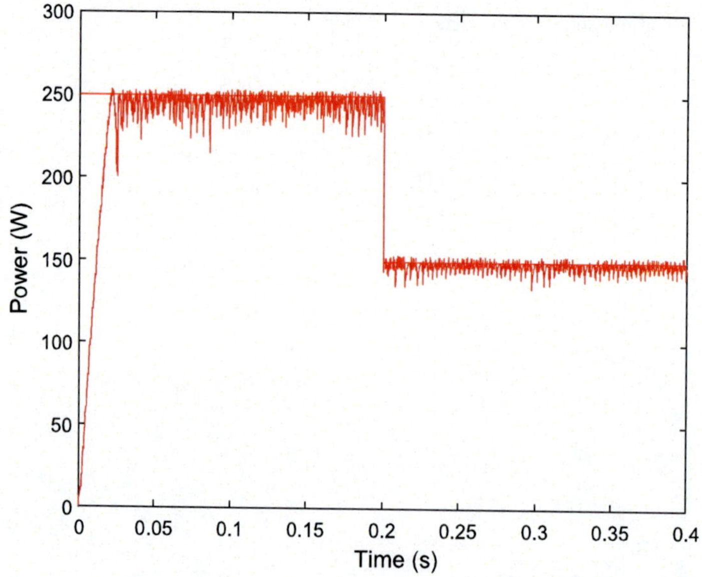

Fig. 11 The evolution of output powers for irradiation varying from 1000 to 600 W / m^2

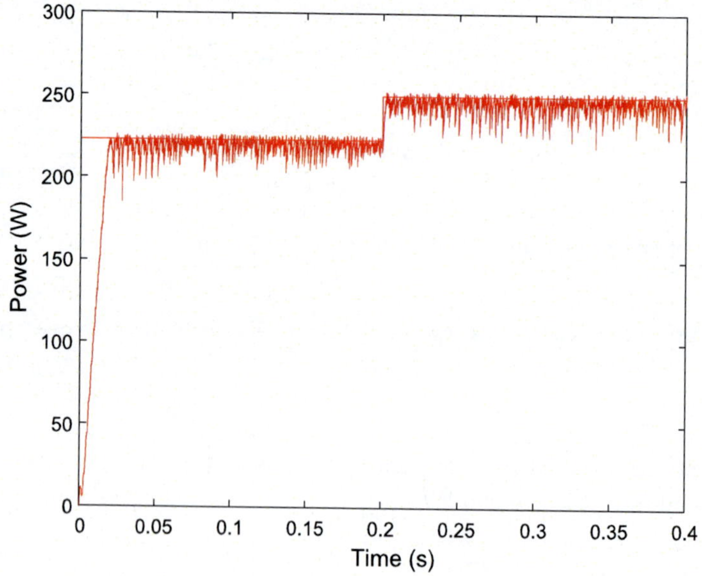

Fig. 12 The evolution of output powers for a temperature varying from 50 to 25 °C

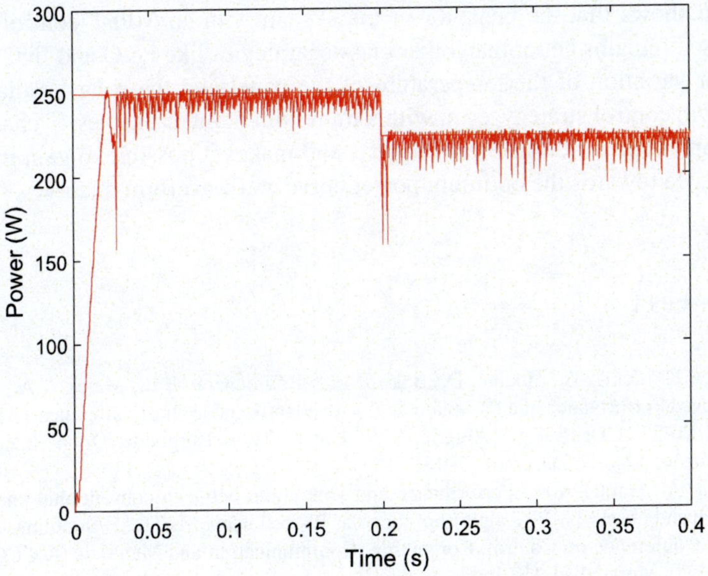

Fig.13 The evolution of output powers for a temperature varying from 25 to 50 °C

Table 2 Table of parameters

PV power	250 W
V_{pm}	31 V
I_{SC}	8.55 A
C	2 μF
L	0.05 mH
Switching frequency	10 kHz

- Fourth case

The fourth case treated is a fixed solar irradiation of 1000 W/m^2, and a temperature incremented from 25 to 50 °C.

Figure 13 represents the response based on the algorithm using fuzzy logic. The behavior using fuzzy logic has fewer ripples and allows an agile response to the system.

Fuzzy logic, therefore, makes it possible to gain in terms of the speed of convergence of the system and the quality of the signal [12, 13].

7 Conclusion

In conclusion, the stability of the whole system is determined by its behavior regarding the parameters change, which is demonstrated in the figures above. This approach shows that the behavior of the system with an MPPT control based on fuzzy logic remains better than other classic strategies like P&O and INC and especially for variation of the temperature of the panel and the solar irradiation. The fuzzy logic control strategy, even with a minimum number of rules, is characterized by an improvement for different requests and makes it possible to gain in speed of convergence towards the optimum power curve and waveform quality.

References

1. Amara, K., Fekik, A., Hocine, D., Bakir, M.L., Bourennane, E.B., Malek, T.A., Malek, A.: Improved Performance of a PV solar panel with adaptive neuro fuzzy inference system ANFIS based MPPT, CDER Alger Algérie, LE2I University of Burgundy, Dijon France, LATEE TiziOuzou, Algéria, December 2018
2. Nigam, A., Gupta, A.K.: Performance and simulation between conventional and improved perturb and observe MPPT algorithm for solar PV cell using MATLAB/Simulink. In: International Conference on Control, Computing, Communication and Materials (ICCCCM), 2016, University Moradabad, UP, India
3. Abbes, H., Abid, H., Loukil, K., Toumi, A., Abid, M.: Etude comparative de cinq algorithmes de commande MPPT pour un système photovoltaïque. In : Conférence Internationale des Energies Renouvelables (CIER'13), 2013 x, Sfax Tunisia
4. Das, S.K., Panda, P.C., Samal, S.: Grid connected PV system with fuzzy logic controller. IEEE, ICCSP (April 2017), February, 2018, Chennai India
5. Lokanadham, M., Bhaskar, K.V.: Incremental conductance based maximum power point tracking (MPPT) for photovoltaic system. Int. J. Eng. Res. Appl. (IJERA) (2012). India
6. Djalab, A., Rezaoui, M.M., Teta, A., Boudiaf, M.: Analysis of MPPT methods: P&O, INC and fuzzy logic (CLF)for à PV system. In: Applied Automation and Industrial Diagnostics Laboratory, Djelfa University, July 2019, Algeria
7. Kandoussi, Z., Boulghasoul, Z., Elbacha, A., Tajer, A.: Fuzzy sliding mode observer based sensorless indirect FOC for IM drives. In: The 3rd World Conference on Complex Systems (WCCS), 2015, Marrakesh, Morocco
8. Prasad, C.B., Sonam, S.K., Reddy, B.R.G., Harika, P.: A fuzzy logic based MPPT method for solar powergeneration. In: International Conference on Intelligent Computing and Control Systems ICICCS, 2017, India
9. Sankar, R., Velladurai, S., Rajarajan, R., Thulasi, J.A.: Maximum power extraction in PV system using fuzzy logic and dual MPPT control. In: International Conference on Energy, Communication, Data Analytics and Soft Computing (ICECDS) (2017)
10. Saleh, A., Azmi, K.F., Hardianto, T.: Comparison of MPPT Fuzzy logic controller based on perturb and observe (P&O) and incremental conductance (InC) algorithm onbuck-boost converter. 2nd International Conference on Electrical Engineering and Informatics (ICon EEI 2018), Batam, Oct. 2018, Indonesia
11. Ziane, A., Necaibia, A., Mostfaoui, M., Bouraiou, A., Sahouane, N., Dabou, R.: A fuzzy logic MPPT for three phase grid connected PV inverter. In: 20th International Middle East Power Systems Conference (MEPCON 2018), Cairo University, Egypt
12. Kechiche, O.B.H.B., Barkaoui, B., Hamza, M., Sammouda, H.: Simulation and comparison of P&O and FuzzyLogic MPPT techniques at diffèrent irradiation conditions. Laboratoire d'énergie et de matérieux, Univ. De Sousse, 2017, Tunisia

13. Bibhu, P., Rosalin, P., Sibani, D., Sibani, G.: Analytical study of MPPT based PV system using fuzzy logic controller. Department of Electrical Engineering, Indira Gandhi institute of technology, Sarang, Dhenkanal, June 2018, India
14. Yousef, A.M., El-Saady, G., Abu-Elyouser, F.K.: Fuzzy logic controller for a photovoltaic array system to ac grid connected. Electrical Engineering Department, Assiut University, Egypt
15. Hannan, M.A., Ghani, Z.A., Mohamed, A., Uddin, M.N.: Real time testing of a fuzzy logic controller-based Grid connected photovoltaic inverter system. Department of EE&SE Universiti Kebangsaan, Malaysia, Bangi, Dept. of ECE Universitite knikal Malaysia, Melaka, Dept. of EE, Lakehead Canada

Artificial Intelligence for Forecasting the Photovoltaic Energy Production

Azeddine El-Hammouchi, Mohammed Bouafia, Nabil El Akchioui, and Amine El Fathi

Abstract Incoming solar energy projections are becoming increasingly relevant because of the significant increase in solar energy generation. The management and operation of photovoltaic solar power plants with energy storage systems necessitates accurate production forecasts, in particular, the ability to predict the amount of energy produced in a given time interval, as well as the occurrence of intermittence that necessitates the use of energy stored in batteries or a conventional emergency source. This paper provides a method for estimating photovoltaic power using artificial neural networks, namely long short-term memory (LSTM). In the realm of deep learning, LSTM is an artificial recurrent neural network design. To discover the best answer, LSTM neural networks may learn from both present and prior information. Furthermore, LSTM cells may learn which data needs to be read, saved, and wiped from memory. Because of these properties, LSTM networks are ideal for projecting photovoltaic energy generation. We picked the Python programming language to program our LSTM neural network since it is particularly powerful for situations that demand a huge database. Our LSTM network had only one input and output at first. The power history is the input, and the expected power is the output. Then we added two additional inputs: irradiation and temperature. In this scenario, our network has three inputs and one output, which is the expected power. Both of these networks have demonstrated good forecasting performance, with the second having a higher precision.

Keywords Artificial intelligence · Neural networks · Recurrent neural networks · LSTM neural networks

1 Introduction

Today, energy production in general and electricity production, in particular, is considered to be representative indicator of a country's economic development.

A. El-Hammouchi (✉) · M. Bouafia · N. El Akchioui · A. El Fathi
LRSDI Laboratory, Faculty of Sciences and Techniques of Al Hoceima, Abdelmalek Essaadi University, Larache, Morocco
e-mail: azeddine.elhammouchi@etu.uae.ac.ma

M. Bendaoud et al. (eds.), *The Proceedings of the International Conference on Electrical Systems & Automation*, https://doi.org/10.1007/978-981-19-0035-8_4

Making energy production planning essential to ensure long-term electricity supply [1]. Unlike traditional energy sources, the production of photovoltaic energy cannot be exactly planned, as the amount of energy produced partly renewable sources depend heavily on external factors such as weather conditions [2].

There are two main methods for the prediction of photovoltaic energy: The first option is to model the photovoltaic system using analytical equations. The majority of efforts are usually focused on establishing accurate irradiance projections, as this is the most important component in energy generation. The physical or parametric method is the name given to this procedure [3, 4]. The second strategy, on the other hand, is to use statistical and machine learning algorithms to directly forecast the output photovoltaic power [5, 6].

The prediction of photovoltaic energy generation by artificial intelligence, specifically artificial neural networks implemented in the Python programming language, will be of interest in this research. Most artificial intelligence-based prediction models take into account a variety of variables that determine the amount of energy produced. These data are used to estimate output power up to a specific time horizon by reducing prediction error (that is, the difference between the predicted value and the actual value) [2].

2 Neural Networks

The most extensively utilized machine learning approaches in solar energy prediction are artificial neural networks. They are based on the way biological neurons work, in which a set of neurons is connected to form a neural network. The learning phase determines the ultimate value of the numerical weights assigned to the synapses between neurons [7, 8].

The learning system of the human brain is assimilated by an artificial neural network. With less computational effort, it can find a relationship between inputs and outputs for linear and nonlinear systems. This set of tools has proven to be effective in a wide range of scenarios and with a big number of inputs. There are numerous topologies to choose from [9].

An artificial neural network is made up of neurons, which are simple processing units connected by a large number of weighted links. The knowledge is saved, and signals or information can pass through.

In order to produce predictions, artificial neural networks require a database that represents the network input values. In order to estimate future values at the network's output, these input numbers are subjected to mathematical calculations [10].

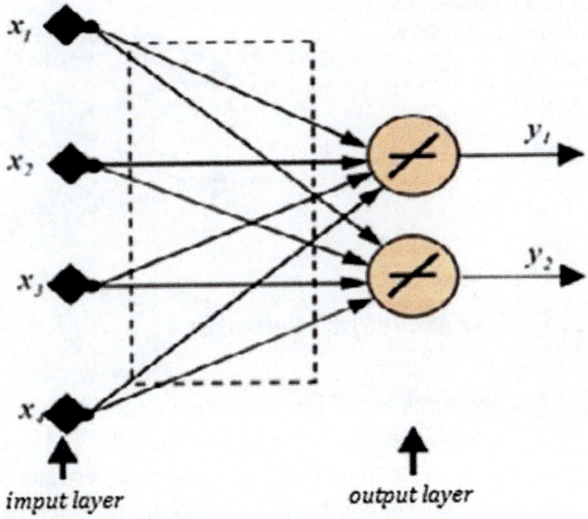

Fig. 1 Architecture of feedforward artificial neural network

2.1 Feedforward Neural Networks

An artificial neural network is a feedforward if it allows signals to flow in only one direction: from inputs to outputs without backtracking. In other words, the output values never jump back to the inputs because there is no backpropagation (loops). The graph of a feedforward artificial neural network is then acyclic (Fig. 1).

2.2 Recurrent Neural Networks

Recurrent artificial neural networks, unlike feedforward artificial neural networks, can send signals in both directions by incorporating loops. These networks become cyclical as a result. Because the presence of cycles has a significant impact on the network's learning capacity and performance, this sort of artificial neural network is more powerful than feedforward neural networks. Recurrent neural networks can handle a wide range of input data sizes. As a result, these are neural networks that propagate impulses in both directions. [11] (Fig. 2).

Recurrent artificial neural networks are dynamic, and their state evolves over time until they reach a stable state. They stay in balance until the input changes and a new equilibrium is established [12].

Recent research has focused on demonstrating the effectiveness of recurrent artificial neural networks in modeling nonlinear dynamical systems. Furthermore, even in the presence of measurement noise, recurrent artificial neural networks may produce long-range predictions and offer greater flexibility in filtering out noisy inputs [13].

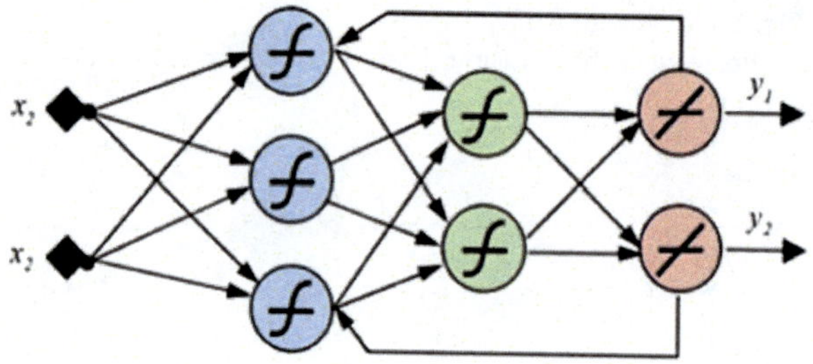

Fig. 2 Architecture of recurrent artificial neural network

2.3 LSTM Neural Networks

LSTM networks are special recurrent neural networks. They are proposed by Sepp Hochreiter and Jürgen Schmidhuber in 1997 [14]. A recurrent neural network is a neural network with recurring connections between neurons. This allows it to learn current and previous information in order to find a better solution [14].

However, when two cells of the recurrent neural network are far apart from each other, it is difficult to obtain useful information due to the problems of disappearance and explosion of the gradient [14]. The solution to this is special neurons called LSTM memory cells [14]. LSTMs are capable of storing useful information over an arbitrary period of time [15]. In addition, LSTM cells have the ability to learn what data is to be read, stored and erased from memory by adjusting three different control gates, namely: $f(t)$ forget gate (to erase unnecessary data), input gate $i(t)$ (entry gate to add new information) and output gate $o(t)$ (exit gate which gives the state of the cell in a moment) as shown in the following figure. LSTM networks can model temporal changes in data and improve forecast results. They are often widely used to predict numerical values such as photovoltaic power (Fig. 3).

An LSTM network has the ability to remove or add information about the state of the cell, carefully regulated by structures called gates. The gates in an LSTM

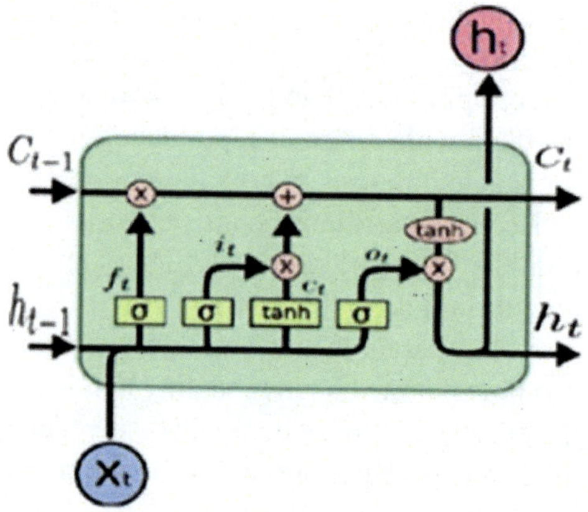

Fig. 3 Structure of a long short-term memory cell

Fig. 4 Forget gate

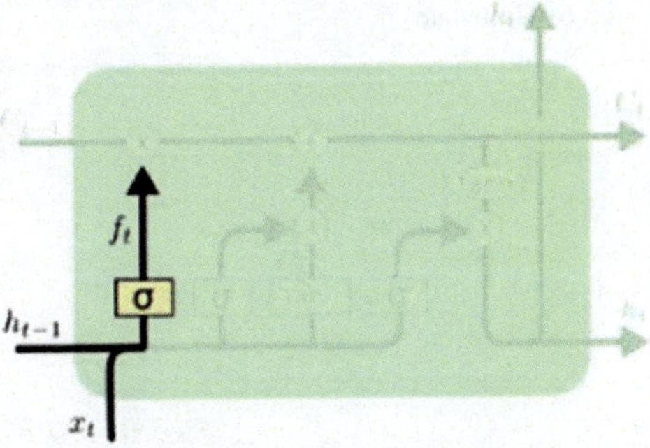

network are analog in the form of a sigmoid, which means they range from 0 to 1. Being analog allows them to backpropagate with them.

a. **Forget Gate**

This gate determines whether information should be kept or discarded: the previous hidden state's information is concatenated with the input data. The numbers between 0 and 1 are then normalized using a sigmoid function. If the sigmoid's output is close to 0, the information must be forgotten, if it is close to 1, the information must be memorized (Fig. 4).

The forget gate's output is given by:

$$f = \sigma(W_F[h(t-1), X(t)] + b_f \quad (1)$$

b. **Input Gate**

This gate's job is to extract information from the already available data. As seen in the following diagram, two functions, the sigmoid function and the hyperbolic tangent function, are applied in parallel (Fig. 5).

The Sigmoid function will then return a vector with a point near 0 indicating that the coordinate in the concatenated vector's equivalent location is unimportant. A coordinate that is near to 1 will, on the other hand, be considered significant.

The output of the sigmoid function is given by the equation below:

Fig. 5 Input gate

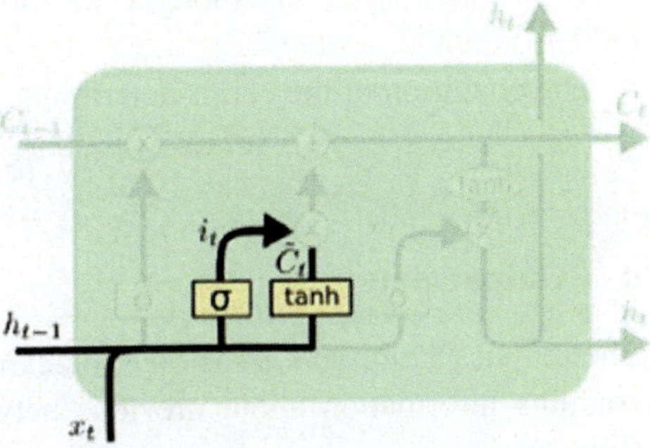

Fig. 6 Cell state

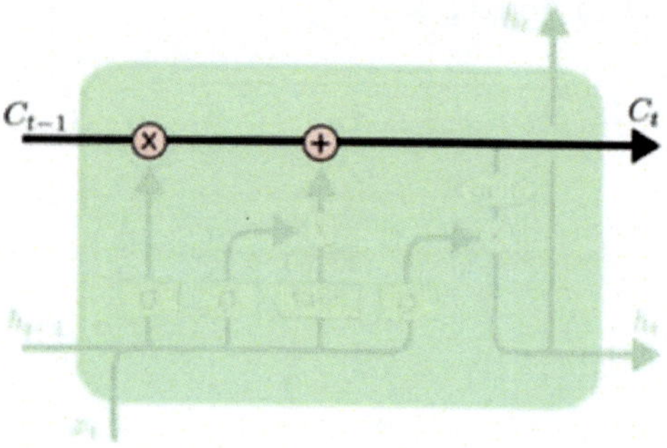

$$i(t) = \sigma(W_i[h(t-1), X(t)] + b_i \tag{2}$$

To avoid overloading the computer with calculations, the tanh will simply normalize (overwrite) the values between -1 and 1. The tanh layer generates a vector of new candidate values $\overline{C}_t$, that can be used to update the state.

$$\overline{C}_t = \tanh(W_C * [h_{t-1}, x_t] + b_C \tag{3}$$

As a result, the product of the two functions will allow you to maintain only the most important data, with the rest being nearly replaced by 0.

c. Cell State

The state of the cell, represented by the horizontal line running across the top of the diagram in the diagram below, is the key to LSTM networks. With only a few modest linear interactions, the cell's state follows the whole chain. It is quite simple for data to pass throughout the chain without being altered (Fig. 6).

We discuss the cell's status before moving on to the last gate (output gate), because the value calculated here is employed there.

The state of the cell may be computed very easily using the forget gate and the input gate: first, we multiply the output with the former state of the cell coordinate by coordinate. This permits you to forget about some data from the prior state that isn't relevant to the new prediction. Then we combine everything (coordinate by coordinate) with the output of the input gate, allowing us to record in the cell's state what the LSTM network thought meaningful (among the inputs and the previous hidden state).

The condition of the cell is determined by:

$$C_t = f_t * C_{t-1} + i_t * \overline{C}_t \tag{4}$$

d. Output Gate

The output gate of an LSTM determines what will be the next hidden state, which contains information about previous network entries and is used for predictions (Fig. 7).

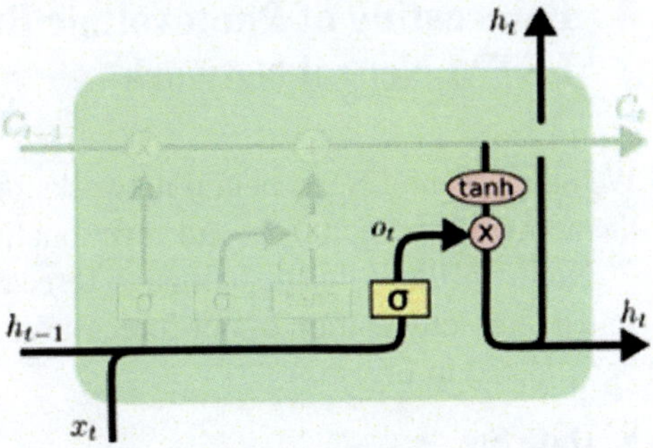

Fig. 7 Output gate

The hyperbolic tangent normalizes the cell's new state, which was determined between −1 and 1 shortly before. The vector formed by concatenating the current entry with the previous hidden state, on the other hand, is passed through a sigmoid function whose goal is to determine which information to keep (close to 0 means that we forget, and close to 1 that we will keep this coordinate of the state of the cell).

$$h(t) = o(t) * \tanh(c(t)) \tag{5}$$

$$o(t) = \sigma(W_o[h(t-1), X(t)] + b_o \tag{6}$$

3 Performance Evaluation

There are two parameters that can be used to assess the efficacy of artificial neural networks in predicting photovoltaic energy generation.

- **The Mean Squared Error (MSE)**

The arithmetic mean of the squares of the differences between the forecast and the actual value is used to calculate this factor. The RMSE factor is calculated as follows:

$$\text{MSE} = \frac{1}{N}\sum_{i=1}^{N}(A_i - F_i)^2 \tag{7}$$

- **The Root Mean Squared Error**

The mean square error (RMSE) is another approach to evaluating predictive quality. The RMSE can be calculated using the equation below:

$$\text{RMSE} = \sqrt{\text{MSE}} \tag{8}$$

4 Forecasting of Photovoltaic Energy Production Using LSTM Neural Network

We detailed the LSTM neural networks in the preceding section by detailing all of the network's components and functionalities.

Now, we'll use this technique to forecast photovoltaic energy production.

The implementation of our forecast for the production of photovoltaic power can be reflected in Fig. 8.

- **Data Set**

We require historical solar power and meteorological data to develop an efficient LSTM neural network model that allows the LSTM network to run and train in order to extract fresh photovoltaic power values that represent the expected power.

The history of power data and meteorological data was obtained from a photovoltaic plant located in Morocco in the city of Assa Zag.

Power (in kW), irradiance (in W/m^2), and temperature (°C) data from this plant were collected with a 5 min step from January 7, 2015 to April 30, 2015.

- **Forecasting by the History of Photovoltaic Power**

We'll start by forecasting PV power based solely on historical power values. The outcomes are depicted in Fig. 9.

Fig. 8 Forecast flowchart

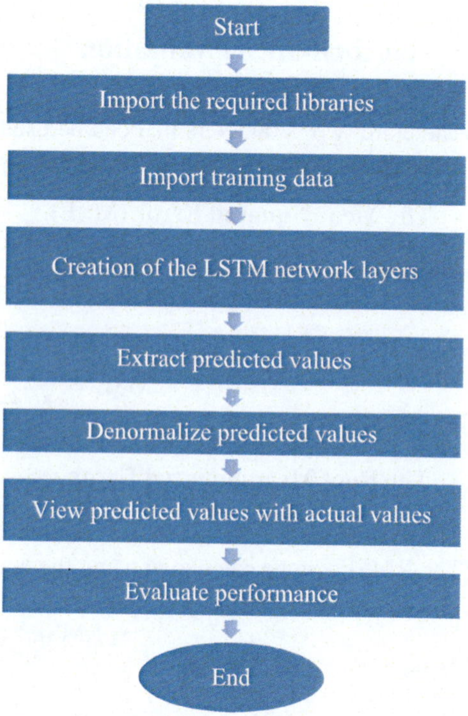

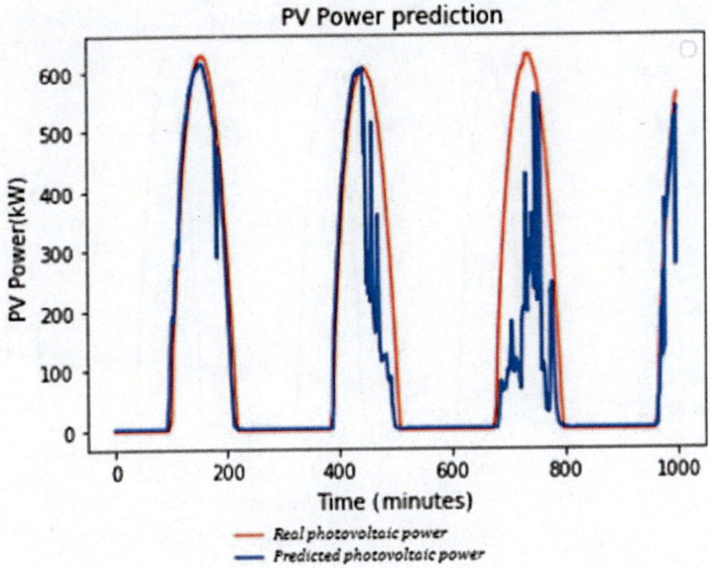

Fig. 9 Forecasting using one input

The MSE and RMSE factors of the forecast based on historical power data are as follows:

$$MSE = 17312.223 \tag{9}$$

$$RMSE = 131.578 \tag{10}$$

Our model's performance evaluation factors are relatively low. Furthermore, the difference between the prediction and actual values is modest, indicating that our LSTM model is effective for this type of forecasting.

Additional variables, such as meteorological data, could be added to increase the quality of our forecast as much as possible.

- **Prediction by Historical Power and Irradiance**

We will now add the irradiance values to the LSTM neural network's inputs in order to increase accuracy. Figure 10 depicts the results achieved.

The following are the performance factors:

$$MSE = 358.299 \tag{11}$$

$$RMSE = 19.629 \tag{12}$$

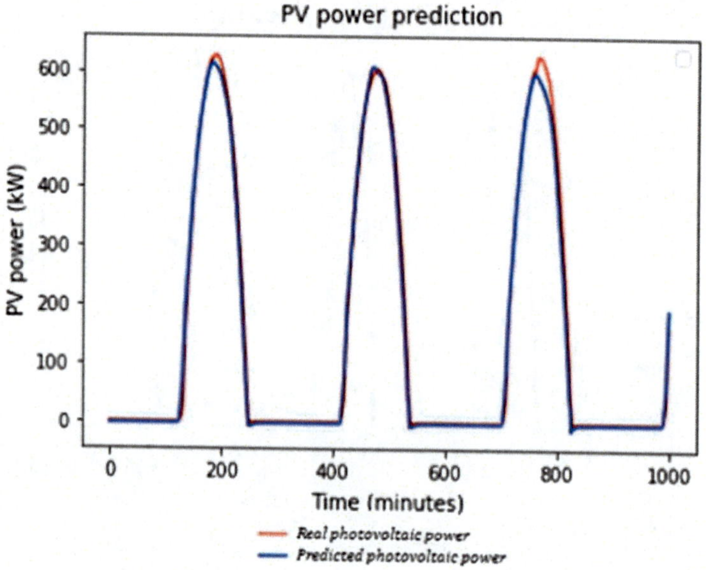

Fig. 10 Forecasting using two inputs

In comparison to the previous scenario, it is apparent that accuracy has become more crucial and errors have been reduced.

- **Forecast by Historical Power, Irradiance and Temperature**

Finally, the temperature values were added to the network's input. Figure 11 depicts the outcomes that we have obtained:

In this scenario, the forecast's MSE and RMSE factors are:

$$MSE = 333.673 \tag{13}$$

$$RMSE = 18.266 \tag{14}$$

Adding another representative variable (temperature) to our forecast improved its accuracy. As a result, it is obvious that combining meteorological data with power history data improves forecast accuracy.

- **Comparison of Results and Discussion**

In our work, we used the Python programming language to develop our LSTM neural network to predict the photovoltaic power generation of the city of Assa Zag.

At first, we tried to create an ideal network architecture in order to obtain good forecasting results. After several tests, we found that the most ideal LSTM network for our situation consisted of four hidden layers, each with fifty neurons. Table 1 displays the results we discovered.

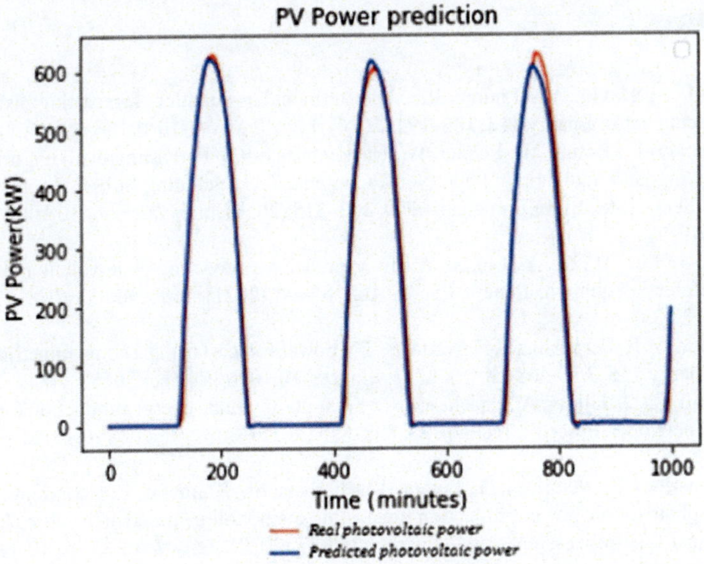

Fig. 11 Forecasting using three inputs

Table 1 MSE and RMSE performance factors based on the number of forecast entries

	One input	Two inputs	Three inputs
MSE	17,312.223	358.299	333.673
RMSE	131.578	19.629	18.266

The inclusion of several factors definitely aids our LSTM neural network in improving prediction accuracy and minimizing the error between real and expected power, as evidenced by the three forecasts we've made. Furthermore, the accuracy of prediction is greatly influenced by the quality of training and test data.

5 Conclusion

The objective of this work is the prediction of the production of photovoltaic energy using artificial intelligence including neural networks of the LSTM type and based on the energy produced in the past as well as meteorological data.

First, we predicted the power based on historical photovoltaic power values. And, in order to improve this forecast, we've included meteorological data, which has proven to be quite useful in enhancing the forecast's quality and accuracy.

References

1. Boland, J., David, M., Lauret, P.: Short term solar radiation forecasting: island versus continental sites. Energy **113**, 186–192 (2016). https://doi.org/10.1016/j.energy.2016.06.139
2. Dannecker, L., Böhm, M., Lehner, W., Hackenbroich, G.: Forcasting evolving time series of energy demand and supply. Lect. Notes Comput. Sci. (including Subser. Lect. Notes Artif. Intell. Lect. Notes Bioinformatics). **6909**, 302–315 (2011). https://doi.org/10.1007/978-3-642-23737-9_22
3. Wang, S., Luo, H., Ge, Y., Liu, S.: A new approach for modeling photovoltaic modules based on difference equation. Renew. Energy **168**, 85–96 (2021). https://doi.org/10.1016/j.renene.2020.12.023
4. Almeida, M.P., Perpiñán, O., Narvarte, L.: PV power forecast using a nonparametric PV model. Sol. Energy **115**, 354–368 (2015). https://doi.org/10.1016/j.solener.2015.03.006
5. Dumitru, C.-D., Gligor, A., Enachescu, C.: Solar photovoltaic energy production forecast using neural networks. Procedia Technol. **22**, 808–815 (2016). https://doi.org/10.1016/j.protcy.2016.01.053
6. Theocharides, S., Makrides, G., Livera, A., Theristis, M., Kaimakis, P., Georghiou, G.E.: Day-ahead photovoltaic power production forecasting methodology based on machine learning and statistical post-processing. Appl. Energy. **268**(4) (2020). https://doi.org/10.1016/j.apenergy.2020.115023
7. Jha, S.K., Bilalovic, J., Jha, A., Patel, N., Zhang, H.: Renewable energy: present research and future scope of Artificial Intelligence. Renew. Sustain. Energy Rev. **77**(4), 297–317 (2017). https://doi.org/10.1016/j.rser.2017.04.018
8. Brunette, E.S., Flemmer, R.C., Flemmer, C.L.: A review of artificial intelligence. ICARA 2009 4th International Conference on Autonomous Robots and Agents, pp. 385–392 (2009). https://doi.org/10.1109/ICARA.2000.4804025
9. Saberian, A., Hizam, H., Radzi, M.A.M., Ab Kadir, M.Z.A., Mirzaei, M.: Modelling and prediction of photovoltaic power output using artificial neural networks. Int. J. Photoenergy **2014** (2014). https://doi.org/10.1155/2014/469701
10. Ahmad, T., Zhang, H., Yan, B.: A review on renewable energy and electricity requirement forecasting models for smart grid and buildings. Sustain. Cities Soc. **55**, 102052 (2020). https://doi.org/10.1016/j.scs.2020.102052
11. Lipton, Z.C., Berkowitz, J., Elkan, C.: A Critical Review of Recurrent Neural Networks for Sequence Learning, pp. 1–38, 2015, [Online]. Available: http://arxiv.org/abs/1506.00019
12. Reddy, B.K., Delen, D.: Predicting hospital readmission for lupus patients: An RNN-LSTM-based deep-learning methodology. Comput. Biol. Med. **101**, 199–209 (2018). https://doi.org/10.1016/j.compbiomed.2018.08.029
13. Toha, S.F., Tokhi, M.O.: MLP and Elman recurrent neural network modelling for the TRMS no. October, 2008. https://doi.org/10.1109/UKRICIS.2008.4798969
14. Sherstinsky, A.: Fundamentals of recurrent neural network (RNN) and long short-term memory (LSTM) network. Phys. D Nonlinear Phenom. **404**, 132306 (2020). https://doi.org/10.1016/j.physd.2019.132306
15. AlKandari, M., Ahmad, I.: Solar power generation forecasting using ensemble approach based on deep learning and statistical methods. Appl. Comput. Informatics (2019). https://doi.org/10.1016/j.aci.2019.11.002

Diagnosis and Classification of Photovoltaic Panel Defects Based on a Hybrid Intelligent Method

Saliha Sebbane, Noamane Ncir, and Nabil El Akchioui

Abstract To enhance the efficiency of the energy generated by a photovoltaic system (PV), a control and monitoring system must be included in the PV system to guarantee that faults are recognized instantly. With the appearance of artificial intelligence-based methods, including machine learning (K-nearest neighbor (k-NN), Decision Tree (DT), Support Vector Machine (SVM), and Artificial Neural Network (ANN)). And through the evaluation of these methods in the classification of photovoltaic faults, the results show that the ANN performs better than other machine learning approaches on the classification of solar field defects. However, due to the artificial neural network's slow learning phase convergence, this article proposes a hybrid diagnostic method based on particle swarm optimization and the neural network (PSO-ANN) to improve the accuracy and the convergence speed of an ANN. To compare the performance of ANNs with the PSO-ANN method, the solar generator's current Ipv and voltage Vpv characteristics are used as identification parameters.

Keywords Photovoltaic system · Fault classification · Artificial intelligence (AI) · Artificial neural network (ANN) · Particle swarm optimization (PSO)

1 Introduction

To ensure a sustainable source of energy, Morocco has adopted a national strategy on renewable energies to reduce the use of fossil fuels. The strategy includes increasing solar energy potential to 5000 MWc in 2030 [1].

S. Sebbane (✉) · N. Ncir · N. El Akchioui
Faculty of Sciences and Technology, University Abdelmalek Essaadi/ Al Hoceima, Tetouan, Morocco
e-mail: saliha.sebbane@etu.uae.ac.ma

N. Ncir
e-mail: noamane.ncir@etu.uae.ac.ma

N. El Akchioui
e-mail: n.elakchioui@uae.ac.ma

M. Bendaoud et al. (eds.), *The Proceedings of the International Conference on Electrical Systems & Automation*, https://doi.org/10.1007/978-981-19-0035-8_5

This policy was accompanied by the intense use of photovoltaic systems (PV) at the national level because of its many advantages. However, PV installations are sudden from time to time failures resulting in a decrease in the electricity produced, which is why a lot of research is devoted to developing diagnostic techniques for PV fields to guarantee reliable And efficient power output.

The diagnostic methods are divided into two categories, the first category contains conventional methods described in literature and industry [2–5], and the second category contains methods based on artificial intelligence (AI) [6]. The research results prove that the artificial neural network (ANN) is the best solution to overcome the limitations of traditional methods and other machine learning methods in the accuracy of detection and identification of defects [7]. However, there is a problem during the use of ANNs to detect and classify faults, such as the slow convergence in the learning phase, which requires performance optimization of ANN.

In this study, the artificial neural network was optimized using a metaheuristic algorithm, to improve diagnostic accuracy and minimize learning time. It is a hybrid diagnostic method based on particle swarm optimization (PSO) and neural network (PSO-ANN). In this article, the first part presents the studied PV field defects, the second and third parts are about the details of the ANN and PSO methods, and the fourth part deals with the combination between the ANN and the PSO method. Finally, the results obtained by ANN are discussed and compared with those of PSO-ANN.

2 Defects of a Photovoltaic Field

A change in the operating conditions of the PV array indicates implicitly that a fault has occurred. This fault can be divided into three categories [8]: physical faults can be a cracking or degradation of photovoltaic modules, such as corrosion and oxidation, the second category are electrical faults which are: open-circuit, short-circuit, and environmental faults include shading and dirt caused by accumulated dust, bird drops and snow.

In this article, the classification is done on three faults which are: partial shading fault, short-circuit fault, and open-circuit fault, as long as these three faults commonly appear and occur in the PV field.

2.1 Partial Shading

Partial shading is the act of obstructing solar radiation from reaching a part of the photovoltaic module contained in a PV field. While a part of the photovoltaic module is exposed to partial shading, and the other part is fully exposed to solar radiation, the output current of the photovoltaic generator is reduced, which leads to the reduction of the total power produced by the PV field.

2.2 *Short-Circuit Fault*

The short-circuit fault produced in a PV field is mainly due to the infiltration of water in the modules, bad wiring between the module and the inverter, or the aging of the PV modules, due to the functioning in long-term of PV system [9].

When a PV module is short-circuited, the voltage is zero. The current in the field becomes equal to the maximum current produced by the modules, and the short-circuited path carries the excess current.

2.3 *Open-Circuit Fault*

An open-circuit fault occurs due to a break in the connection wires between the PV cells. This type of fault is usually caused by the poor quality of the connections between PV cells, through the manufacturing process, especially when one of these high resistance connections occurs.

3 Artificial Neural Network (ANN)

Artificial neural networks are mathematical models inspired by biology. As the basic rebels of these networks, artificial neurons were originally the result of the hope to model the function of biological neurons, which can be divided into three major entities: a cell body, a set of dendrites, and an axon [10].

Perceptrons also called artificial neurons or formal neurons, are designed to mimic the functions of biological neurons [11]. Therefore, there are several levels of abstraction, depending on the precision of the modeling as shown in Fig. 1.

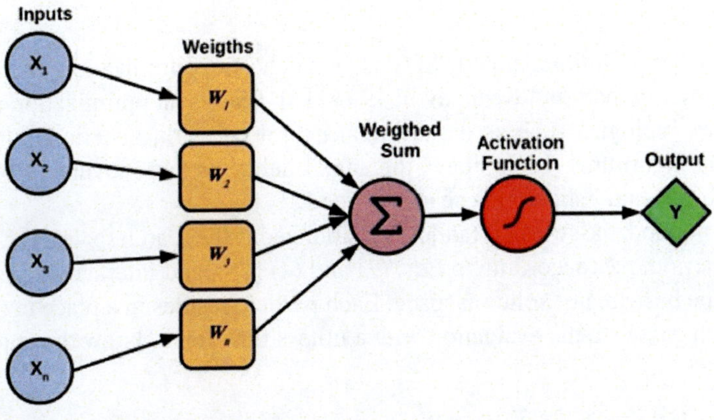

Fig. 1 Representation of an artificial neuron

From the figure above, we will consider the following entities:

- Entries labeled in vector form, representing dendrites.
- A noted output representing the axon.
- The parameters noted w and b affect neuronal function.

This representation can be modeled as an Eq. (1) that shows how to calculate the output by multiplying the input by the weight then adding them to the bias. Finally, the result of the summation goes through the transfer function f, which is generally nonlinear.

$$\hat{y} = f(\langle \omega, x \rangle + b) \tag{1}$$

The growing complexity of the problem makes it very difficult to resolve. That is why a single neuron cannot cope with complex problems. So the effective method that allows solving such a problem is to assemble several perceptrons to obtain what is called the multilayer perceptron (MLP), which is an early-acting ANN and can be applied in several applications such as recognition of images, shapes or speech, prediction, etc. [12].

MLP is suitable for fixed-size data, such as images, and can contain at least three main layers, the input layer, the hidden layer, and the output layer. In other words, information flows from the input layer to the output layer via the hidden layer to decrease the error between the desired output and the measured one. This error is calculated with Eq. (2).

$$J(a) = \frac{1}{2m} \sum_{i=1}^{m} (y_i - \hat{y}_i)^2 \tag{2}$$

4 Particle Swarm Optimization (PSO)

Particle Swarm Optimization (PSO) is a stochastic optimization metaheuristic proposed by Eberhart and Kennedy in 1995 [13]. PSO is an optimization algorithm inspired by biologies such as the artificial neural network, genetic algorithms, or ant colony algorithms. It simulates the social behavior and movement of animals (insects, birds, and fish, etc.) in search of food [14].

In this method, the swarm is randomly initialized in the search space. The members of the swarm move (according to Eqs. (3) and (4) [15], and interact with each other to reach the best area of solution space. Each particle resides in a place in the search space, then passes to the evaluation with a fitness function to know the quality of its position.

$$V(t+1) = V(t) + C_1 r_1 (\text{Pb}(t) - X((t)) + C_2 r_2 (\text{Pg}(t) - X(t)) \tag{3}$$

1. Initialize initial position Xi, initial velocity Vi, Pb, Pg ;
2. Generate random particles (P) ;
3. for each particle (Pi)
4. Calculate the new velocity using the equation (3);
5. Calculate a new position using the equation (4);
6. Calculate the fitness value of each particle (Pi);
7. If f(Pi) is better than f(pb) then
8. Pb = Pi;
9. If f(Pb) is better than f(Pg) then
10. Pg = pb;
11. End for

Fig. 2 Pseudo code of PSO algorithm

$$X(t+1) = X(t) + V(t+1) \tag{4}$$

X the position of the particle in the search space;
V the velocity of the particle;
Pb the position of the best solution through which the particle has passed;
Pg the position of the best known solution of the whole swarm;
C_1 and C_2 acceleration coefficients;
r_1 and r_2 two random numbers in the interval [0,1].

The conventional algorithm of the PSO method begins with the initialization of the initial population (N), position, and velocity of movement. The particles move in each iteration using Eqs. (3) and (4), and the fitness function of each particle in the swarm is calculated to indicate the best position of the whole population (Pg).

After the evaluation of the fitness function value, Pb and Pg are updated, this procedure is repeated until the stop criterion is reached. The pseudo-code of the PSO is cited in Fig. 2 [16].

5 PSO-ANN

5.1 Design of the PSO-ANN Hybrid Method

ANNs are characterized by the efficiency of detection and diagnosis of almost all faults affecting the PV field [17]. ANN effectiveness refers to the value of weights and biases, but when using the ANN-based diagnostic model, convergence is slow in the training step.

This study proposes a hybrid diagnostic method based on particle swarm optimization and artificial neural network (PSO-ANN). The goal of using PSO is to optimize the ANN in terms of convergence and accuracy.

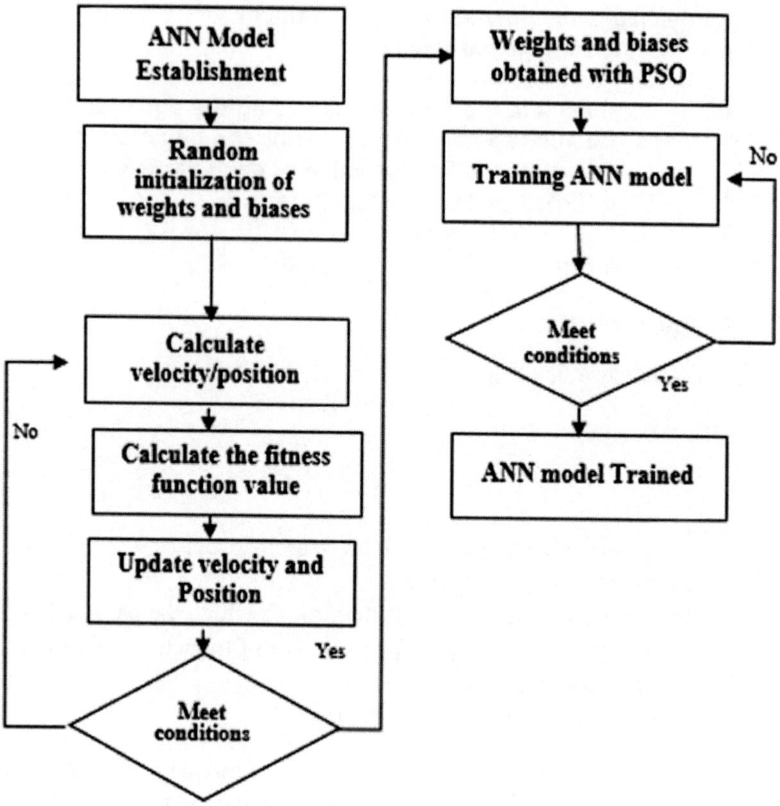

Fig. 3 Flowchart of the PSO-ANN hybrid method

In the PSO-ANN model, each position contains the initial weights and biases, weights and biases are optimized before the learning phase, then used to build the network, and next optimized again in the training step by minimizing the objective function. Figure 3 shows the flowchart of the PSO-ANN model.

5.2 *Methodology*

The methodology adopted in the present study is composed of two steps: the first step involves collecting the data used to train the neural network, while the second step determines the neural network structure used for defect classification. To collect the data necessary for training the neural network, a PV field (4 modules in series, 4 in parallel) is simulated on MATLAB/Simulink, under standard test conditions, and different fault conditions presented previously.

The data assembled in this process for fault identification are the current Ipv and the voltage Vpv generated by the PV field, including 331,315 samples.

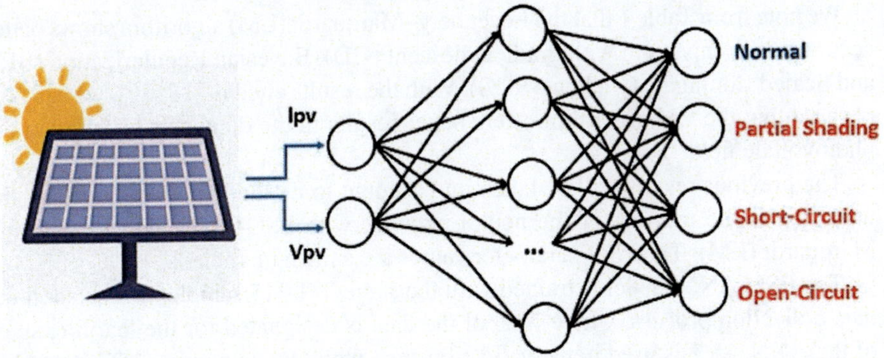

Fig. 4 Schematic diagram of proposed PSO-ANN method to detect the state of PV Field

Concerning the neural network structure used in the diagnostic model, we used a 3-layer neural network, the input layer contains two neurons that receive the current Ipv and the voltage Vpv, the hidden layer containing ten neurons, finally, the output layer composed of four neurons, each one presents the state of the PV field, such as a normal functioning or a type of fault (partial shading fault, short-circuit fault and open circuit fault). Figure 4 shows the simulation model diagram used to identify the state of the PV field.

5.3 *Result and Discussion*

In this article, we performed a neural network-based diagnostic model. The neural network structure chosen in this work is the MLP with three layers. The latter was trained first with four training algorithms, including Gradient Descent (GD), Levenberg–Marquardt (LM), Bayesian Regularization (BR), and Scaled Conjugate Gradient (SCG), to determine which is the correct one, which gives the small error compared to other algorithms. We have chosen the sigmoid tangent function as the activation function to obtain the output results as probabilities between -1 and 1. The objective function used in the diagnostic model and the mean squared error (MSE) (shown in Eq. 2). The results of training the ANN with the four algorithms are given in Table 1.

Table 1 The results of training the neural network with the four training algorithms

Algorithm	Iterations number	Error (MSE)	Accuracy (%)
GD	183	0.0945	90.55
LM	449	0.0010	99.89
BR	639	0.0013	99.87
SCG	957	0.0371	96.29

We note from Table 1 that the Levenberg–Marquardt (LM) algorithm shows better accuracy in comparison with gradient descent (GD), Bayesian Regularization (BR), and Scaled Conjugate Gradient (SCG). With the results cited in [18–20], we can also observe that the LM algorithm offers better results in the diagnosis of faults in the photovoltaic field.

The previous results will be taken into account to fix the learning algorithm, the next step is the training of the neural network with PSO-ANN with Levenberg–Marquardt (LM). The PSO parameter values are shown in Table 2.

The PSO-ANN model is trained with the same 331 315 samples collected in the data collection process, where 70% of the data is designated for the learning stage of the ANN, 15% is used in the validation step, while the remaining 15% is used in the learning test step. The simulation result is shown in Fig. 5.

From the analysis of Fig. 5, we notice that the PSO algorithm plays a crucial role in the performance optimization of ANN. With the PSO-ANN model, the learning step is accomplished with less iteration (365) in comparison with the ANN model (449), which means that the PSO-ANN converges faster than the ANN model. The

Table 2 The values of the important parameters of PSO

Parameters	Symbol	Value
Population	N	25
Max iteration	Tmax	100
Cognitive factor	C_1	1.5
Social factor	C_2	2.5

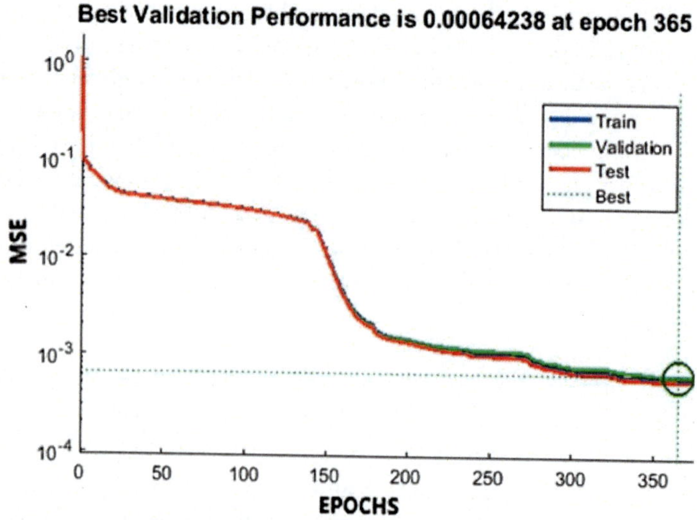

Fig. 5 The mean squared error of PSO-ANN model

Table 3 The classification results of the PSO-ANN model

State of PV field	Scenarios	Predicted result			
		N	PS	SC	OC
Partial shading (PS)	Irradiation of the first column = 800 W/m^2	0.0199	**1**	−0.001	−0.001
Normal (N)	Under standard test condition Irradiance = 1000 W/m^2	**1**	−0.037	0.005	−0.005
Short-circuit (SC)	Two modules are short-circuited	−0.0159	0.020	**0.9966**	−0.0013
Open-circuit (OC)	Four modules are disconnected	0.115	−0.106	−0.006	**1**
Open-circuit (OC)	Six modules are disconnected	0.030	0.146	0.046	**0.7773**
Short-circuit (SC)	Only one module are short-circuited	0.008	−0.010	**1**	0.000
Partial Shading (PS)	Irradiation of the first line = 500 W/m^2 and Irradiation of three modules in the first column = 500 W/m^2	−0.0009	**0.66**	0.039	0.296

classification precision of the hybrid method reaches a value of 99.94% though the model precision ANN reaches just 99.89%.

Reference [21] proposed a meta-heuristic method based on the PSO and the neural network for diagnosing the photovoltaic field. The results of this reference and the results obtained in our article prove that the diagnostic model obtained in our research is more precise and the training of the model requires only two inputs and necessitates fewer populations and fewer iterations for the parameters of the PSO algorithm.

The trained model based on PSO-ANN is tested under several states of the chosen PV field, the states are assembled between the three defects and the case of normal functioning. The results are shown in Table 3.

6 Conclusion

In this research, we have combined the metaheuristic optimization algorithm PSO and the artificial neural network ANN. This combination aims to reduce the convergence time of the artificial neural network and thus improve the accuracy of the classification.

The proposed PSO-ANN model is trained with current Ipv and voltage Vpv. These parameters are obtained by simulating the PV field using "Matlab Simulink" in several cases. The results show that the PSO-ANN model achieves an accuracy of 99.94% in fewer iterations than the ANN-based diagnostic model.

The diagnostic model based on the hybrid method was tested on a PV field under several faults to determine its capacity. The results prove that the fault identification is done correctly with good precision, which offers an accurate diagnostic model.

Acknowledgements We acknowledge the financial support for this research from the "Centre national pour la Recherche Scientifique et technique", CNRST, Morocco.

References

1. Stratégie Enérgetique Nationale Horizon: 2019–10–08, p.264, Ministére de l'Energie, des Mines de l'Eau et de l'Environnement (2030)
2. Bun, L.: Détection et localisation de défauts dans un système photovoltaïque, p. 207 (2011)
3. Hajji, M. et al.: Multivariate feature extraction based supervised machine learning for fault detection and diagnosis in photovoltaic systems. Euro. J. Control S0947358019304054 (2020). https://doi.org/10.1016/j.ejcon.2020.03.004
4. Appiah, A.Y., Zhang, X., Ayawli, B.B.K., Kyeremeh, F.: Review and Performance Evaluation of Photovoltaic Array Fault Detection and Diagnosis Techniques. Int. J. Photoenergy. **2019**, e6953530 (2019). https://doi.org/10.1155/2019/6953530
5. Spataru, S., Sera, D., Kerekes, T. and Teodorescu, R.: Diagnostic method for photovoltaic systems based on light I–V measurements. Solar Energy **119**, 29–44 (2015). https://doi.org/10.1016/j.solener.2015.06.020
6. Mellit, A., Tina, G.M. and Kalogirou, S.A.: Fault detection and diagnosis methods for photovoltaic systems: a review. Renew. Sustain. Energy Rev. **91**, 1-17 (2018). https://doi.org/10.1016/j.rser.2018.03.062
7. Pahwa, K., Sharma, M., Saggu, M.S., Mandpura, A.K.: Performance evaluation of machine learning techniques for fault detection and classification in PV array systems. In: 2020 7th International Conference on Signal Processing and Integrated Networks (SPIN), Noida, India, pp. 791–796 (2020). https://doi.org/10.1109/SPIN48934.2020.9071223
8. Basnet, B., Chun, H., Bang, J.: An intelligent fault detection model for fault detection in photovoltaic systems. J. Sens. **2020**, e6960328 (2020). https://doi.org/10.1155/2020/6960328
9. Garoudja, E., Harrou, F., Sun, Y., Kara, K., Chouder, A., Silvestre, S.: Statistical fault detection in photovoltaic systems. Solar Energy. **150**, 485–499 (2017). https://doi.org/10.1016/j.solener.2017.04.043
10. Ghritlahre, H.K., Prasad, R.K.: Application of ANN technique to predict the performance of solar collector systems—a review. Renew. Sustain. Energy Rev. **84**, 75–88 (2018). https://doi.org/10.1016/j.rser.2018.01.001
11. Hussain, M., Dhimish, M., Titarenko, S., Mather, P.: Artificial neural network based photovoltaic fault detection algorithm integrating two bi-directional input parameters. Renew. Energy **155**, 1272–1292 (2020). https://doi.org/10.1016/j.renene.2020.04.023
12. Cortés, B., Sánchez, R.T., Flores, J.J.: Characterization of a polycrystalline photovoltaic cell using artificial neural networks. Solar Energy. **196**, 157–167 (2020). https://doi.org/10.1016/j.solener.2019.12.012
13. Kennedy, J., Eberhart, R.: Particle Swarm Optimization, p. 7 (1995)
14. ParticleSwarm Optimization (PSO).: Data Analytics Post. https://dataanalyticspost.com
15. /Lexique/particle-swarm-optimization-pso/
16. Zemzami, M., Elhami, N., Makhloufi, A., Itmi, M., Hmina, N.: Application d'un modèle parallèle de la méthode PSO au problème de transport d'électricité. IncertFia. 17(1) (2017). https://doi.org/10.21494/ISTE.OP.2017.0127

17. Dereli, S., Köker, R.: Strengthening the PSO algorithm with a new technique inspired by the golf game and solving the complex engineering problem. Complex Intell. Syst. **7**(3), 1515–1526 (2021). https://doi.org/10.1007/s40747-021-00292-2
18. Li, B., Delpha, C., Diallo, D., Migan-Dubois, A.: Application of artificial neural networks to photovoltaic fault detection and diagnosis: a review. Renew. Sustain. Energy Rev. **138**, 110512 (2021). https://doi.org/10.1016/j.rser.2020.110512
19. Sabri, N., Tlemçani, A.: Faults diagnosis in stand-alone photovoltaic system using artificial neural network. In: 2018 6th International Conference on Control Engineering and Information Technology (CEIT), Istanbul, Turkey, 2018, pp. 1–6. https://doi.org/10.1109/CEIT.2018.8751924
20. Laarabi, B., Tzuc, O.M., Dahlioui, D., Bassam, A., Flota-Bañuelos, M., Barhdadi, A.: Artificial neural network modeling and sensitivity analysis for soiling effects on photovoltaic panels in Morocco. Superlattices Microstruct. **127**, 139–150 (2019). https://doi.org/10.1016/j.spmi.2017.12.037
21. Chine, W., Mellit, A., Lughi, V., Malek, A., Sulligoi, G., Massi Pavan, A.: A novel fault diagnosis technique for photovoltaic systems based on artificial neural networks. Renew. Energy **90**, 501–512. 2016. https://doi.org/10.1016/j.renene.2016.01.036
22. Liao, Z., Wang, D., Tang, L., Ren, J., Liu, Z.: A heuristic diagnostic method for a PV system: triple-layered particle swarm optimization–back-propagation neural network. Energies **10**(2), 226 (2017). https://doi.org/10.3390/en10020226

New Reduced Form Approach and an Efficient Analytical Model for the Prediction of the Five Parameters of PV Generators Under Non-STC Conditions

Kawtar Tifidat and Noureddine Maouhoub

Abstract In this work, a new numerical method is presented to estimate the five parameters' values of the single-diode model of photovoltaic (PV) modules operating under standard test conditions (STC). The prediction is done with high precision and without using any approximations. The current method is based on the reduction of the research space from five to two unknowns. The other three parameters are calculated analytically using the two numerically extracted parameters. Moreover, the new approach is based only on the remarkable points under standard test conditions available on the datasheet, so it does not require any kind of measured current–voltage characteristics of the PV panels. As the second stage, this paper introduces also a new contribution to the transfer of the five parameters from STC to non-standard test conditions (non-STC). In order to prove the effectiveness of the technic presented in this work, the method was applied to a PV generator, for which it showed a high accuracy compared with other introduced methods in the literature.

Keywords Photovoltaic module · Single-diode model · Reduced form · Maximum power point · Module temperature · I-V characteristics · Parameter extraction

1 Introduction

The increasing implementation of photovoltaic systems makes the modeling of these systems a great requirement, in order to be able to predict their performances and the parameters influencing them. For this purpose, various equivalent circuits were used in the literature to model a photovoltaic generator. But, thanks to its high simplicity-precision ratio, the single-diode equivalent model of the five parameters (The photo-generated current I_{ph}, the reverse saturation I_s, the series resistance R_s, the parallel resistance R_p, and the ideality factor n) modeling the physical effects inside the PV generator remains the most used [1–3].

K. Tifidat · N. Maouhoub (✉)
Laboratory of Electronics, Signal Processing and Physical Modeling, Department of Physics, Faculty of Sciences, Ibn Zohr University, B.P 8106, Agadir, Morocco
e-mail: n.maouhoub@uiz.ac.ma

M. Bendaoud et al. (eds.), *The Proceedings of the International Conference on Electrical Systems & Automation*, https://doi.org/10.1007/978-981-19-0035-8_6

According to the literature, several methods have been utilized for the estimation of the single-diode model's five parameters allowing the prediction of the maximum power that a PV generator can provide, under varying levels of illumination (E) and temperature (T). Therefore, the used methods can be divided into three kinds of parameters' extraction. In the first group, we have the works using analytical formulas, which rely on the values of the key-points available on the PV panel's datasheet [2–4]. In the second group, we find all numerical methods exploiting the measured characteristics and trying to minimize the errors between the experimental and calculated I–V curves [5–11]. As a third group, we cite the methods based on the meta-heuristic algorithms corresponding to natural phenomena to estimate the five parameters and predict the maximum power point [12, 13].

Villalva's method [8] stay one of the most cited works in this area, it is based on a three remarkable-points curve adjustment technique and uses an iterative process to estimate the series resistance's value. Therefore, it can be concluded that this method ensures an accurate prediction only in the vicinity of the maximum power point and not for the whole I–V curve. In addition, the iterative process may increase the calculation cost as well. Unlike Villalva's method which fixes the value of the ideality factor in advance, Benahmida et al. [6] propose an iterative method for which the calculation of this value is done using the available values of the three remarkable points on the generator's datasheet. The mean inconvenient of both methods is the use of the iterative process that may increase the needed time for the identification, and also the use of the error minimization only for the maximum power point which can lead to the loss of accuracy for the estimation of the rest of the I–V curve. Maouhoub [11] introduces an analytical approach based on the least-squares technic for characterizing the PV panels. But this method necessitates the calculation of the slope in the vicinity of the short-circuit point to calculate the value of R_s. Then, the value of R_s will be very influenced by the precision of the measurement of this slope. Nassar-eddine et al. [9] compare two different extraction methods, the first is an analytical approach, and the second is iterative, also based on the minimization of the absolute error only for the maximum power point.

With the aim of identifying PV generators' parameters with high precision under different weather conditions, a new numerical method is introduced in this work. The extraction starts first by calculating the values of n and R_s solving a two nonlinear equations' system, instead using an iterative process which can increase the calculation time and lead to the loss of precision. The three values of the other remaining parameters (R_p, I_s, and I_{ph}) are directly calculated using three analytical equations which do not rely on any kind of approximations. Furthermore, in order to be able to model a photovoltaic generator working at external conditions using its identification already done for standard test conditions, a new contribution has been made in the current paper.

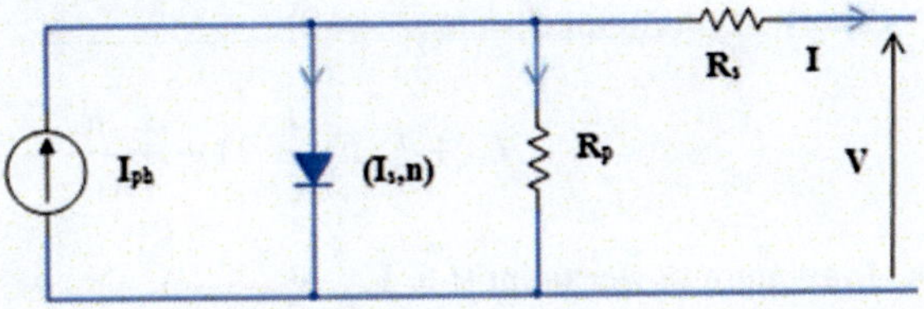

Fig. 1 Single-diode electrical model of a solar cell

2 Single-Diode Modelling

The current method uses the single-diode circuit shown in Fig. 1 as a basic model for the identification. In this equivalent circuit, I_{ph} is the photo-generated current, I_s and n are respectively the reverse saturation current and the ideality factor, of the diode modeling the semi-conductor material that the solar cell contains, and R_p and R_s are respectively the shunt resistance and the series resistance.

The equation linking the current between the two surfaces of the PV module to its output voltage is given as a function of the five parameters as follow [2–9]:

$$I = I_{ph} - I_s\left(\exp\left(\frac{V + IR_s}{nC_1}\right) - 1\right) - \frac{V + IR_s}{R_p} \tag{1}$$

$C_1 = N_s V_{th}$, where N_s is the number of cells connected in series, forming the PV generator. V_{th} correspond to the thermal voltage giving by: $V_{th} = \frac{K_B T}{q}$. K_B is the constant of Boltzmann equals to $1.38064852.10^{-23}$ J/K, and q is the electron charge equals to $1.\ 60217646.10^{-19}$ C.

Using the Lambert W function, the implicit Eq. (1) can have an analytical solution given by the following formula [14, 15]:

$$I = \frac{I_s + I_{ph}}{1 + R_s G_p} - \frac{G_p}{1 + R_s G_p} V - \frac{nC_1}{R_s} \times W\left(\frac{I_s R_s}{nC_1(1 + R_s G_p)} \exp\left(\frac{V + R_s(I_s + I_{ph})}{nC_1(1 + R_s G_p)}\right)\right) \tag{2}$$

where $G_p = \frac{1}{R_p}$.

3 Method's Theory

With the aim of extracting the expressions that will allow the calculation of the five parameters' values exploiting the four values corresponding to the key-points of the I-V curve at STC, the application of the Eq. (1) to the remarkable points will be requested [3, 6, 9, 11]:

- Short-circuit point ($I = I_{sc}$, $V = 0$)

$$I_{sc} + I_s(E_{sc} - 1) + \frac{I_{sc} R_s}{R_p} - I_{ph} = 0 \tag{3}$$

- Maximum power point ($I = I_{mp}$, $V = V_{mp}$)

$$I_{mp} - I_{ph} + I_s\left(E_{mp} - 1\right) + \frac{V_{mp} + I_{mp} R_s}{R_p} = 0 \tag{4}$$

- Open-circuit point ($I = 0$, $V = V_{oc}$)

$$I_{ph} = I_s(E_{oc} - 1) + \frac{V_{oc}}{R_p} \tag{5}$$

where:

$$E_{sc} = \exp\left(\frac{I_{sc} R_s}{nC_1}\right), E_{mp} = \exp\left(\frac{V_{mp} + I_{mp} R_s}{nC_1}\right) \text{ and } E_{oc} = \exp\left(\frac{V_{oc}}{nC_1}\right) \tag{6}$$

The series resistance and the ideality factor's estimation require two equations containing just R_s and n as parameters to be determined. To this end, we use as a first equation the expression giving the fill-factor FF as a function of the four parameters: R_s, n, I_s, and R_p, given by the following formula [16, 17]:

$$FF = \frac{I_{mp}^2}{V_{oc} I_{sc}}\left(R_s + \frac{R_p}{\frac{I_s R_p}{nC_1} E_{mp} + 1}\right) \tag{7}$$

Otherwise, given that FF measures the I–V characteristic's squareness, it is also given as the ratio of the maximum power provided by a real PV generator to the maximum power of an ideal generator, and it can be calculated also as:

$$FF = \frac{V_{mp} I_{mp}}{V_{oc} I_{sc}}$$

To reduce the number of unknown parameters in the Eq. (7), we replace Iph in Eq. (4) by its expression of the Eq. (5), and finally, we get the reverse saturation current Is only as a function of the three parameters R_s, n, and R_p as:

$$I_s = \frac{\frac{V_{oc} - V_{mp} - I_{mp} R_s}{R_p} - I_{mp}}{E_{mp} - E_{oc}} \tag{8}$$

To get rid of I_s from Eq. (7), we replace (8) in (7), so that (7) becomes an expression linking only the three parameters R_s, n, and R_p from which we can get R_p as a function of n and R_s:

$$R_p = \frac{C_2 - \exp(C_2) + 1}{\frac{I_{mp}}{nC_1} + \frac{I_{mp}^2}{(FFV_{oc} - R_s I_{mp}^2)}(1 - \exp(C_2))} \tag{9}$$

where:

$$C_2 = \frac{V_{oc} - V_{mp} - I_{mp} R_s}{nC_1} \tag{10}$$

Finally, to use the "fsolve" function of MATLAB, which solves nonlinear equations' systems, Eq. (7) is considered as the first equation of the two equations' system, which will be used for the numerical extraction of n and R_s. The second equation can be extracted by replacing Iph in Eq. (3) with its expression of the Eq. (5):

$$I_{sc} - I_s(E_{oc} - E_{sc}) - \frac{V_{oc} - I_{sc} R_s}{R_p} = 0 \tag{11}$$

In which I_s and R_p are given respectively in Eqs. (8) and (9) only as function of n and R_s.

To guarantee the rapid convergence of the system's resolution towards the accurate solutions, the right choice of their initial guesses is necessary. To this end, the approximate analytical expression of R_s proposed by Kumar et al. [3], and the formula of n used by Nassar-eddine et al. [9] and Maouhoub [11], given in the system below, are used for the initialization.

$$\begin{cases} n_0 = \dfrac{K_v - \frac{V_{oc}}{T_{STC}}}{C_1\left(\frac{K_i}{I_{sc}} - \frac{3}{T_{STC}} - \frac{E_g}{K_B T_{STC}^2}\right)} \\ R_{s0} = \dfrac{V_{mp}}{I_{mp}} - \dfrac{2V_{mp} - V_{oc}}{(I_{sc} - I_{mp}) \times \ln\left(1 - \frac{I_{mp}}{I_{sc}}\right) + \frac{I_{mp}}{I_{sc} - I_{mp}}} \end{cases} \tag{12}$$

K_i (A/°C) and K_v (V/°C) are respectively the temperature coefficient of short-circuit current, and the temperature coefficient of open-circuit voltage. E_g is the band gap energy.

After the numerical prediction of the values of n and R_s, the values of R_p, I_s, and I_{ph} are calculated respectively using the analytical Eqs. (9), (8), and (5).

4 Results and Discussion

To confirm the accuracy of the current reduced form, the photovoltaic generator Kyocera KC200GT operating under STC (E = 1000 W/m^2 and T = 25 °C) is chosen to apply the method [18]. The measured parameters from the I-V characteristics of the KC200GT PV module working under STC are (I_{sc} = 8.21A; V_{mp} = 26.89 V; I_{mp} = 7.66A; V_{oc} = 33.07 V). As the first task, we calculate the five parameters of

Table 1 Predicted parameters for the multi-crystalline KC200GT operating under STC

	Villalva et al. [8]	Maouhoub [11]	Benahmida et al. [6]	Nassar-Eddine et al. [9]	New method
n	1.0772	1.0805	1.0805	1.0758	1.0982
R_s (Ω)	0.2280	0.1950	0.2229	0.3080	0.2364
R_p (Ω)	195.61	59.8480	186.15	193.04	296.73
I_{ph} (A)	8.2201	8.3232	8.2203	8.2233	8.2170
I_s (nA)	1.9753	2.1031	2.1186	2.1523	3.0451
NRMSE (%)	1.0646	1.6612	1.2052	7.4747	0.9954

the PV generator for STC. Then, we estimate its I-V characteristics for non-STC. The precision of the method can be judged by using several statistical indicators. For this work, we select the Normalized Root Mean Square Error (NRMSE), and the absolute error (AE) [6, 11, 12, 17]:

$$\mathrm{NRMSE} = \frac{\sqrt{\frac{1}{N} \times \sum_{i=1}^{N} \left(I_{i,\mathrm{Measured}} - I_{i,\mathrm{Predicted}}\right)^2}}{\frac{1}{N}\sum_{i=1}^{N} I_{i,\mathrm{Measured}}} \tag{13}$$

$$\mathrm{AE} = \left|X_{i,\mathrm{Measured}} - X_{i,\mathrm{Predicted}}\right| \tag{14}$$

where N represents the number of measured data points, and X symbolizes current I or voltage V.

4.1 *Results for KC200GT Under STC*

Table 1 regroups the predicted values of the five parameters using the new method, compared to four other selected methods from literature. The table contains as well the values of the provided NRMSE for the multi-crystalline KC200GT by different methods. Based on the results, it can be observed that the value of NRMSE supplied by the new approach is the lowest compared to the other estimating technics.

4.2 *Results for KC200GT Under Non-STC*

In order to estimate the values of the five parameters in external conditions and then calculate the I–V characteristics of the PV generator, equations below are employed to make the transfer of I_{ph} and I_s from STC to Non-STC [6, 9, 11]:

$$I_{ph}(E,T) = \left(I_{ph,STC} + K_i(T - T_{STC})\right) \times \frac{E}{E_{STC}} \tag{15}$$

$$I_s(E,T) = \frac{I_{ph}(E,T) - \frac{V_{oc}(E,T)}{R_p(E,T)}}{\exp\left(\frac{V_{oc}(E,T)}{N_s V_{th} n(E,T)}\right) - 1} \tag{16}$$

To extract the other three parameters, we use the expressions used by Petrone, where it is assumed that the three parameters are independent of temperature variations. They only depend on irradiance levels, as the following equations show [19]:

$$R_s(E,T) = R_{s,STC}\left(\frac{I_{sc,STC} V_{oc}(E,T_{STC})}{V_{oc,STC} I_{sc}(E,T_{STC})}\right) \tag{17}$$

$$R_p(E,T) = R_{p,STC}\left(\frac{I_{sc,STC} V_{oc}(E,T_{STC})}{V_{oc,STC} I_{sc}(E,T_{STC})}\right) \tag{18}$$

$$n(E,T) = n_{STC}\left(\frac{V_{oc}(E,T_{STC})}{V_{oc,STC}}\right) \tag{19}$$

where I_{sc} (E,T) is given as [20]:

$$I_{sc}(E,T) = I_{sc,STC} \times \left(\frac{E}{E_{STC}}\right)^{\beta} (1 + K_i(T - T_{STC})) \tag{20}$$

The use of all five equations requires as well a prior knowledge of short-circuit current and open-circuit voltage values in the external conditions. For this purpose, a modified expression of V_{oc} is used to calculate the value of this parameter under non-standard test conditions [2, 17].

$$V_{oc}(E,T) = \left(V_{oc,STC} + K_v(T - T_{STC})\right) \times \left(\frac{E}{E_{STC}}\right)^{\alpha} \tag{21}$$

To demonstrate the suggested model's accuracy, it is compared to two additional models from literature, which are listed below as:

- Model 1 [2, 6]:

$$V_{oc}(E,T) = V_{oc,STC} + K_v(T - T_{STC}) + N_s n V_{th,STC} \ln\left(\frac{E}{E_{STC}}\right) \tag{22}$$

- Model 2 [2]:

$$V_{oc}(E,T) = V_{oc,STC} + K_v(T - T_{STC}) + \gamma(E - E_{STC}) \tag{23}$$

Table 2 Required parameters for switching to non-STC for the multi-crystalline Kyocera KC200GT

Parameter	α	β	γ	K_v (V/°C)	K_i (A/°C)
Value	0.04422	1.005	0.002467	−0.123	0.00318

$V_{oc,STC}$ is the open-circuit voltage at STC. α, β and γ are three adjustment coefficients. Table 2 shows the values of the used coefficients for KC200GT.

Figure 2a presents the measured values of the open-circuit voltage and the calculated values using the introduced modified model and compares them to two other models from the literature for the PV generator KC200GT working under 25 °C and various levels of irradiance. As it is observed, when compared to other models, the new model gives the highest accuracy. The thing that can also be seen from Fig. 2b, which presents the absolute errors between the measured data points and the calculated ones using different models, where the new model gives the lowest values of absolute error for the whole levels of irradiance.

Figure 3a presents measured I-V curves and the calculated ones based on the predicted five parameters' values at STC and using the transfer equations from STC to non-STC. As it can be seen from Fig. 3b showing the calculated values of NRMSE using the new model of Voc and the other two models of literature, the new method provides the best accuracy giving the lowest values of NRMSE for all levels of irradiance and which does not exceed 2.5% in the worst case (Fig. 3b). On the first hand, Fig. 4a presents the predicted I-V curves using the current method as well as the measured curves for the PV panel KC200GT under 1000 W/m^2 and different levels of temperature. On the other hand, Fig. 4b gives the values of NRMSE for various levels of temperature and which do not exceed 3%.

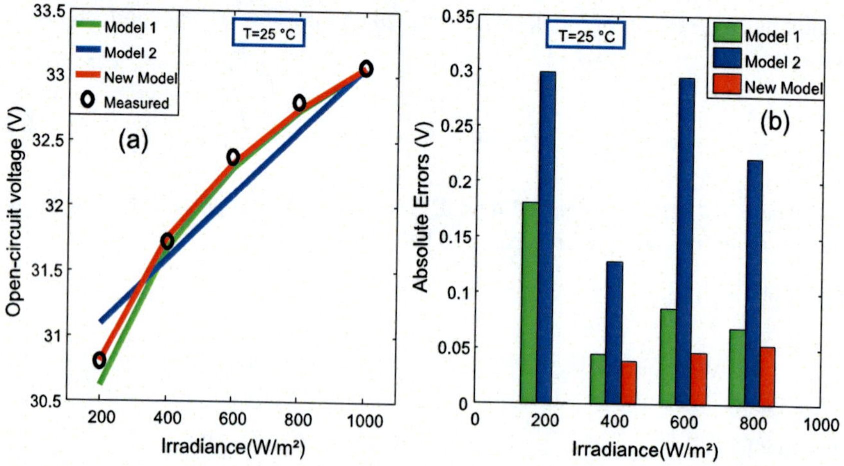

Fig. 2 **a** Measured and calculated open-circuit voltages using the three models for various levels of irradiance. **b** The absolute errors corresponding

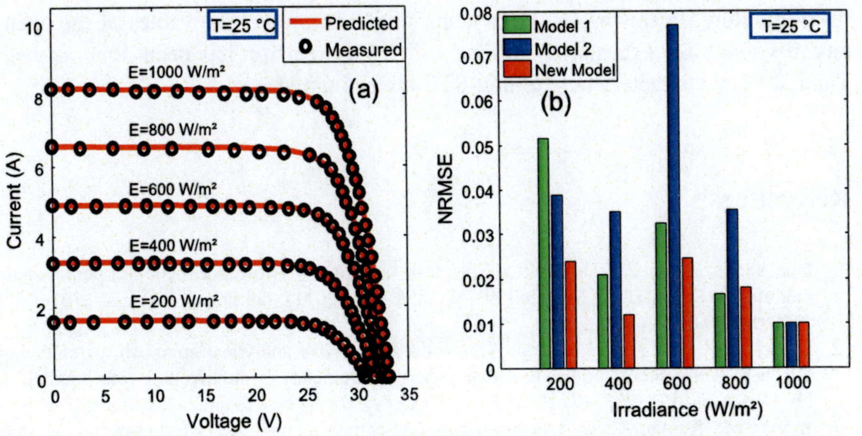

Fig. 3 **a** Measured data and predicted I-V curves at different levels of irradiance. **b** Normalized root mean square errors obtained using the three models of the open-circuit voltage

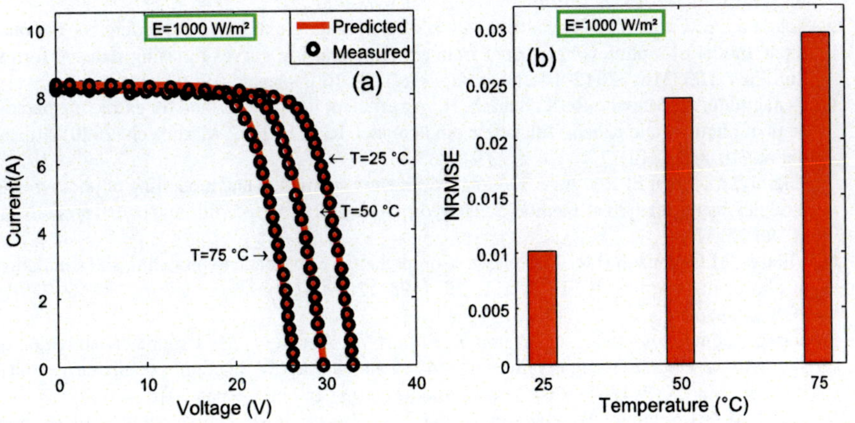

Fig. 4 **a** Predicted I–V curves and measured data at various levels of temperature for the PV generator Kyocera KC200GT. **b** The corresponding normalized root mean square errors

5 Conclusion

This paper proposes an accurate reduced form for estimating the values of the single-diode model's five parameters. The introduced method uses only the provided information about the remarkable-points in the datasheet, and it does not rely on any approximations or iterative processes. That makes the proposed method efficient, simple, and fast. The approach has been tested for the multi-crystalline Kyocera KC200GT and found to be the most accurate compared to the other selected methods from literature. In this paper, a new contribution to predict the values of the open-circuit voltage for external conditions with high precision has been done as well. Then, the I-V characteristics for non-STC are estimated.

References

1. Ishaque, K., Salam, Z., Taheri, H.: Simple, fast, and accurate two-diode model for photovoltaic modules. J. Sol. Mat. &. Sol. Cell. **95**, pp. 856–594 (2011). doi:https://doi.org/10.1016/j.solmat.2010.09.023
2. El Achouby, H., Zaimi, M., Ibral, A., Assaid, E.M.: New analytical approach for modeling effects of temperature and irradiance on physical parameters of photovoltaic solar module. J. En. Con. Man. **177**, 258–271 (2018). https://doi.org/10.1016/j.enconman.2018.09.054
3. Kumar, M., Kumar, A.: An efficient parameters extraction technique of photovoltaic models for performance assessment. Sol. Energy. **158**, 192–206 (2017). https://doi.org/10.1016/2017.09.046
4. Hejri, M., Mokhtari, H., Azizian, M.R., Söder, L.: An analytical-numerical approach for parameter determination of a five-parameter single-diode model of photovoltaic cells and modules. In. J. Sust. En. (2013). https://doi.org/10.1080/14786451.2013.863886
5. Laudani, A., Fulginei, F.R., Salvani, A.: High performing extraction procedure for the one-diode model of a photovoltaic panel from experimental I-V curves by using reduced forms. Sol. Ener. **103**, 316–326 (2014). https://doi.org/10.1016/j.solener.2014.02.014
6. Benahmida, A., Maouhoub, N., Sahsah, H.: An efficient iterative method for extracting parameters of photovoltaic panels with single diode model. R. E. D. E. C. Marrakech (2020). https://doi.org/10.1109/REDEC49234.2020.9163858
7. Elkholy, A., Abou El-Ela, A.A.: Optimal parameters estimation and modelling of photovoltaic modules using analytical method. J. Heliyon. **5**(7) (2019). https://doi.org/10.1016/j.heliyon.2019.e02137
8. Villalva, M.G., Gazoli, J.R., Filho, E.R.: Comprehensive approach to modeling and simulation of photovoltaic arrays. IEEE. T.P.E.L. **25**, 1198–1208 (2009). https://doi.org/10.1109/TPEL.2009.2013862
9. Nassar-eddine, I., Obbadi, A., Errami, Y., El fajri, A., Agunaou., M.: Parameter estimation of photovoltaic modules using iterative method and the Lambert W function: a comparative study. En. Con. Man. **119**, 37–48 (2016). https://doi.org/10.1016/j.enconman.2016.04.030
10. Zaimi, M., El Achouby, H., Zegoudi, O., Ibral, A., Assaid, E.M.: Numerical method and new analytical models for determining temporal changes of model-parameters to predict maximum power and efficiency of PV module operating outdoor under arbitrary conditions. J. En. Con and Man **177**, 258–271 (2020). https://doi.org/10.1016/j.enconman.2018.09.054
11. Maouhoub, N.: Photovoltaic module parameter estimation using an analytical approach and least squares method. J. Com. Elec. 784–790 (2018). https://doi.org/10.1007/s10825-017-1121-5
12. Ridha, H.M., Gomes, C., Hizam, H.: Estimation of photovoltaic module model's parameters using an improved electromagnetic-like algorithm. Neural. Com. Appl. **119**, 37–48 (2020). https://doi.org/10.1007/s00521-020-04714-z
13. Zhang, C., Zhang, Y., Su, J., Gu, T., Yang, M.: Performance prediction of PV modules based on artificial neural network and explicit analytical model. J. Ren. Sust. En. **12**, 013501. (2020). https://doi.org/10.1063/1.5131432
14. Ortiz-Conde, A.J.F., Sánchez, G., Muci, J.: New method to extract the model parameters of solar cells from the explicit analytic solutions of their illuminated I–V characteristics. Sol. En. Sol. Cell. **90**, 352–361 (2006). https://doi.org/10.1016/j.solmat.2005.04.023
15. Pindado, S., Roibás-Millán, E., Cubas, J., Miguel Álvarez, J., Alfonso-Corcuera, D., Cubero-Estalrrich, J.L., Gonzalez-Estrada, A., Sanabria-Pinzón, M., Jado-Puente, R.: Simplified

Lambert W-function math equations when applied to photovoltaic systems modeling. IEEE Trans. Ind. Appl **57**, 1779–1788 (2021). https://doi.org/10.1109/tia.2021.3052858
16. Dadu, M., Kapoor, A., Tripathi, K.N.: Effect of operating current dependent series resistance on the fill factor of a solar cell. Sol. En. Materials. Sol. Cell. **71**, 213–218 (2002). https://doi.org/10.1016/S0927-0248(01)00059-9
17. Tifidat, K., Maouhoub, N., Benahmida, A.: An efficient numerical method and new analytical model for the prediction of the five parameters of photovoltaic generators under non-STC conditions. E3S Web of Conf. **297**, 01034 (2021). https://doi.org/10.1051/e3sconf/202129701034
18. KC200GT High Efficiency Multicrystal Photovoltaic Module Datasheet Kyocera (2000). http://www.kyocera.com.sg/products/solar/pdf/kc200gt.pdf
19. Petrone, G., Ramos-Paja, G. A., Spagnuolo, G.: Photovoltaic sources modeling, pp. 21–44. Wiley (2017). https://doi.org/10.1002/9781118755877.ch2
20. Zhang, C., Zhang, Y., Su, J., Gu, T., Yang, M.: Modeling and Prediction of PV module performance under different operating conditions based on power-Law I-V model. IEEE. J. Photov. 1816–1827 (2020). https://doi.org/10.1109/JPHOTOV.2020.3016607

Parameter Prediction of Solar Cell's Double Diode Model Using Neural Network

Fayrouz Dkhichi

Abstract In order to monitor the behavior of the solar cell, we tried to identify the intrinsic structure of the double diode model of the solar cell for different values of temperature and irradiance. In this context, we tried to predict the values of the seven electrical parameters of solar cell according to the two meteorological factors by using a feedforward artificial neural network. This tool allows a dynamic prediction of the seven parameters. To achieve our goal, we trained our network by the Levenberg–Marquardt algorithm using learning data and we tested its ability of prediction by a test data which are completely different. Therefore, our network determines the seven parameters in an optimal way, such as it gives the appropriate value of each electrical parameter for any value of temperature and irradiance. The obtained results show how each parameter varies according to the two meteorological factors.

Keywords Solar cell · Artificial neural network · Seven intrinsic parameters · Optimization · Hidden neurons · Metrological factors

1 Introduction

Several external factors such as temperature, sand storms, humidity, etc., influence negatively the performances of a PhotoVoltaic (PV) generator over time [1]. The performance degradation of the generator occurs at the basic structures which constitute it, which are the solar cells. Hence the interest in studying the response of these cells. To do this, the researchers proposed several mathematical and electrical models [2–5].

These models present the intrinsic phenomena which govern the PN junction of the solar cell. The obtained response which presents the intrinsic behavior strongly depends on irradiance and temperature. In order to study the influence of these meteorological factors, we need a knowledge model that presents their relationship with

F. Dkhichi (✉)
Department of Electrical Engineering, Network, Computer Science, Telecommunications and Multimedia Laboratory, Superior School of Technology, Hassan II University of Casablanca, Casablanca, Morocco
e-mail: fayrouz.dkhichi@etu.univh2c.ma

M. Bendaoud et al. (eds.), *The Proceedings of the International Conference on Electrical Systems & Automation*, https://doi.org/10.1007/978-981-19-0035-8_7

solar cell behavior. However, in literature, there is not a direct mathematical equation expressing the intrinsic behavior of solar cells with irradiance and temperature. The absence of a knowledge model leads us to think about a black-box model like the artificial neural network [6, 7].

From different solar cell's models in the literature, we choose in this paper to predict the seven parameters of the solar cell's double diodes model [8]. Therefore, we proposed a feedforward neural network based on several hidden neurons. The inputs are defined by the irradiance and temperature and the outputs are presented by the seven electrical parameters. These outputs should be found for different values of temperature and irradiance. The learning of our network is ensured by Levenberg–Marquardt algorithm.

2 Solar Cell's Double Diode Model

The light photons absorbed and converted by a semiconductor to a photocurrent I_{ph} allows to shed light on the solar cell as a current generator. For this reason, the current source is considered as the core of solar cell's electric model. This source is connected in parallel with two diodes (D1 and D2) instead of just one and a shunt resistor R_{sh} [9].

The latter is considered to model the leaks caused by the imperfections of the semiconductor material. On other hand, another resistor is put in series R_s to present the leaks that occurred when electrons pass through the metal contacts [10] (Fig. 1).

In order to study and analyze closely the intrinsic behavior of a solar cell, modeling the solar cell by a mathematical equation is essential [11]:

$$I_{PV} = I_{ph} - I_{s1}\left(\exp\left(\frac{V_{PV} + R_s I_{PV}}{n_1 V_{th}}\right) - 1\right) - I_{s2}\left(\exp\left(\frac{V_{PV} + R_s I_{PV}}{n_2 V_{th}}\right) - 1\right) - \frac{V_{PV} + R_s I_{PV}}{R_{sh}} \quad (1)$$

where

I_{PV} Generated photovoltaic current,

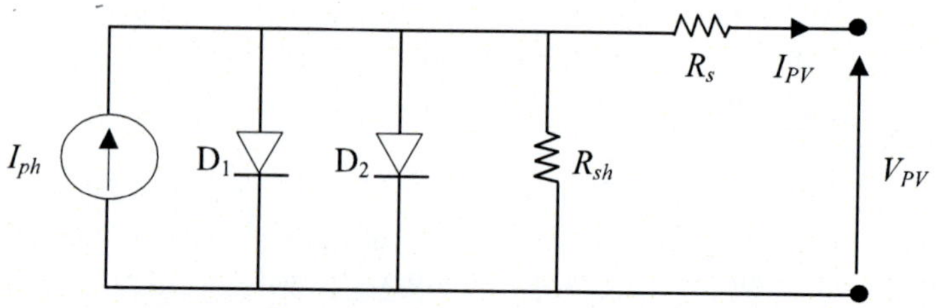

Fig. 1 Equivalent electrical circuit of a solar cell based on two diodes

V_{PV} Generated photovoltaic voltage,
V_{th} Thermal voltage, $V_{th} = AT/q$,
I_{s1} and I_{s2} Saturation currents of the diode 1 and diode 2,
n_1 and n_2 Ideality factors of diode 1 and diode 2,
T Temperature in Kelvin,
A Boltzmann constant $A = 1.38064852 \times 10^{-23}\ m^2\ kg\ s^{-2}\ K^{-1}$,
q Electric load of the electron $q = -1.602 \times 10^{-19}$ C.

A I_{PV} value obtained from Eq. (1) corresponds to a specific value of Temperature (T) and irradiance (G). Therefore, when T and/or G change, I_{PV} also changes. This impact of the meteorological factors on the I_{PV} is explained by the influence of the seven intrinsic parameters on the response of the solar cell.

3 Parameter Prediction for Different Values of G and T

The IPV (VPV) response is generated at the output of the solar cell for an accurate value of temperature (T) and irradiance (G). When the two metrological factors change, the seven values of electrical parameters Rs, Rsh, Iph, Is1, Is2, n1 and n2 change also. Which leads to the variation of the characteristic IPV (VPV). The aim of this work is to predict the values of the seven intrinsic parameters for different values of T and G. this prediction based on the mathematical equation Eq. (1) of the cell seems limited. In order to ensure this identification, it is necessary to apply the equation Eq. (1) several times, which requires time and memory space during the several codes execution. Therefore, a knowledge model is not operational in our case and we need to adopt a faster and more optimal model based on black box context. We cite a frequently used model: Feedforward Artificial Neural Networks (ANN).

4 Artificial Neural Network for Parameter Prediction

4.1 Feedforward Artificial Neural Network Model

In order to predict Rs, Rsh, Iph, Is1, Is2, n1 and n2 parameters according to each value of T and G, we have designed a very suitable neural network (Fig. 2).

where

i Index of output neurons,
j Index of hidden neurons,
m Index of inputs,
w_{ij} Weight connecting the hidden neurons to the output neurons,
w_{jm} Weight connecting the input neurons to the hidden neurons,
b_{ij} Bias of the output layer's neurons,

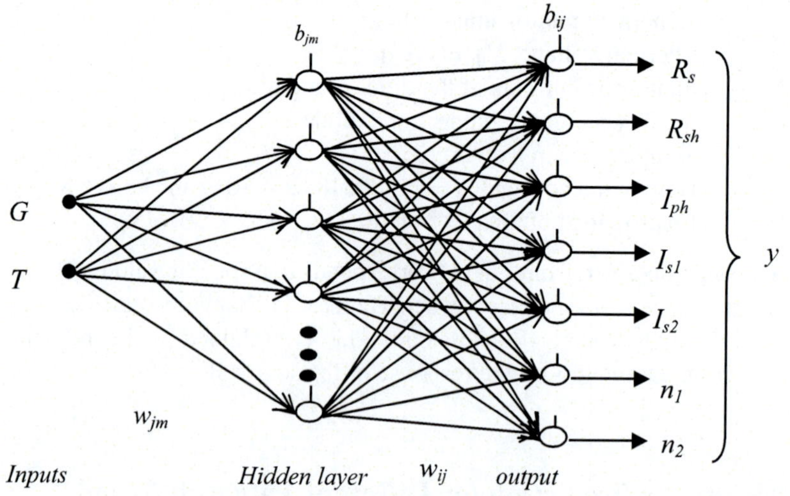

Fig. 2 Architecture of neural network for parameter prediction

b_{jm} Bias of the hidden layer's neurons,
y Networks output.

A neural network is divided into layers: The hidden layer and the output layer. Each layer contains a well-defined number of neurons, to mention 7 neurons in the output layer which presents the seven electrical parameters. The number of neurons in the hidden layer is determined after a further detailed study. The network has two inputs which are the two metrological parameters T and G.

4.2 Learning Process

The network learning is provided by an optimization algorithm of an error function *Mean of Sum Squared Errors* (MSSE) such as [12]:

$$MSSE = \frac{1}{N}\sum_{j=1}^{N}\left[\sum_{i=1}^{Ns}\left[(y_{calculated}(i, j) - y_{measured}(i, j))^2\right]\right] \tag{2}$$

N indicates the number of samples, *j* is the index of used target data, *i* is the index of used output, $y_{calculated}$ are the output values calculated by the network and $y_{measured}$ are the target values of the outputs.

The learning process consists of adjusting the weights *w* of each layer by the Levenberg–Marquardt algorithm, the learning process is represented by the diagram in Fig. 3.

During each adjustment of all weights at each iteration during the iterative learning process, a test phase is triggered. During this phase, another error function is used

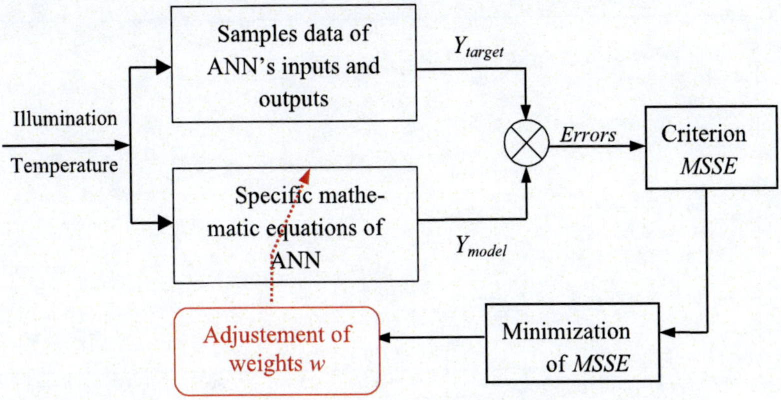

Fig. 3 Block diagram of ANN learning process

called MSSEtest, completely independent of that used in the learning phase. The sample data forming the test set is different from that constituting the training set, in order to ensure that the adjusted w parameters are capable of responding to any sample data.

5 Results and Discussion

In order to train our network efficiently, we used 680 samples of inputs and of outputs. This set is divided into two groups of samples, the first one consists of 70% of samples which present the training set, the second one which is the test set contains 30% of samples [13].

5.1 Impact of the Number of Hidden Neurons on the Network Response

To deduce the configuration of the hidden layer of the neural network of identification of the seven electrical parameters, we tried different numbers of hidden neurons and we concluded the one that corresponds to the minimum value of MSSEtest, obtained for ten successive executions (Fig. 4).

The hidden layer should include 23 hidden neurons. This number allows the neural network the best performance in parameter identification of the seven electrical parameters.

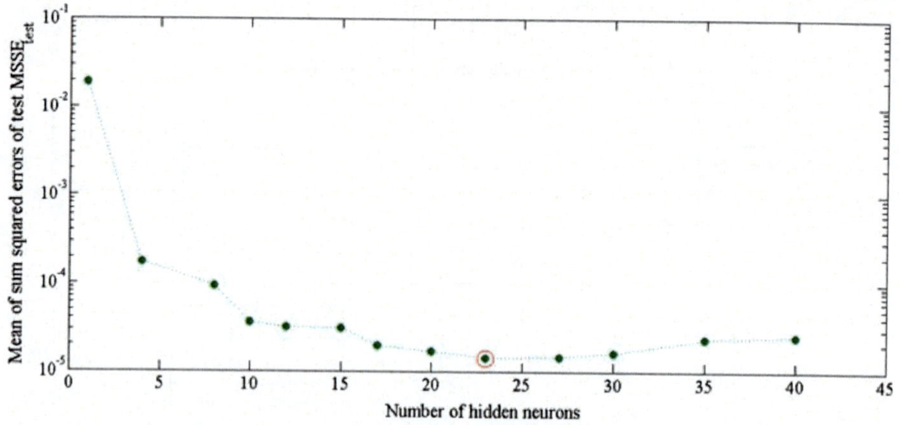

Fig. 4 Evolution of the mean of sum squared errors of the test according to the number of hidden neurons in the identification network of the seven electrical parameters

5.2 Network Behavior During Training

Always by ensuring the learning of the network by the back-propagation algorithm based on Levenberg–Marquardt method, we present in Figs. 5 and 6 the evolution of the learning process of the identification network of the seven electrical parameters as well as the evolution of the process of testing its capacity for generalization.

The training stop is done at iteration 563 by cross-validation using data from the test set. As shown in Fig. 6, just after iteration 563, the MSSEtest error begins to increase more than six times and in multiple times (as marked by blue circles), thus presenting an over-learning of the network.

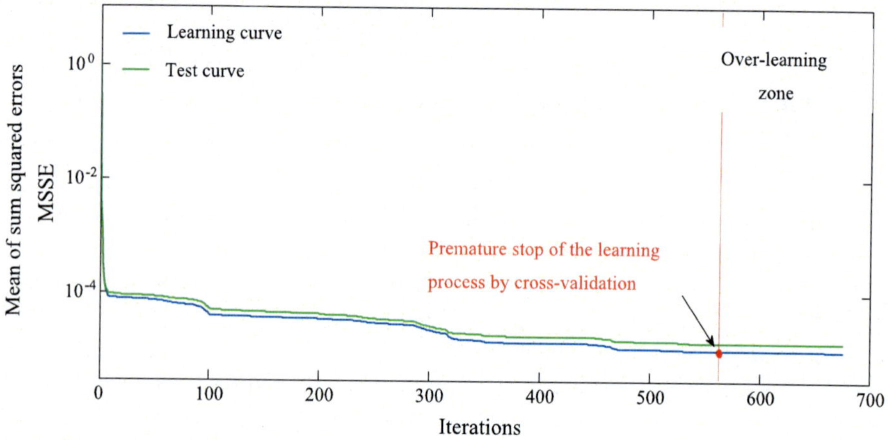

Fig. 5 Evolution of MSSE of learning and test occurred during training of the neural network

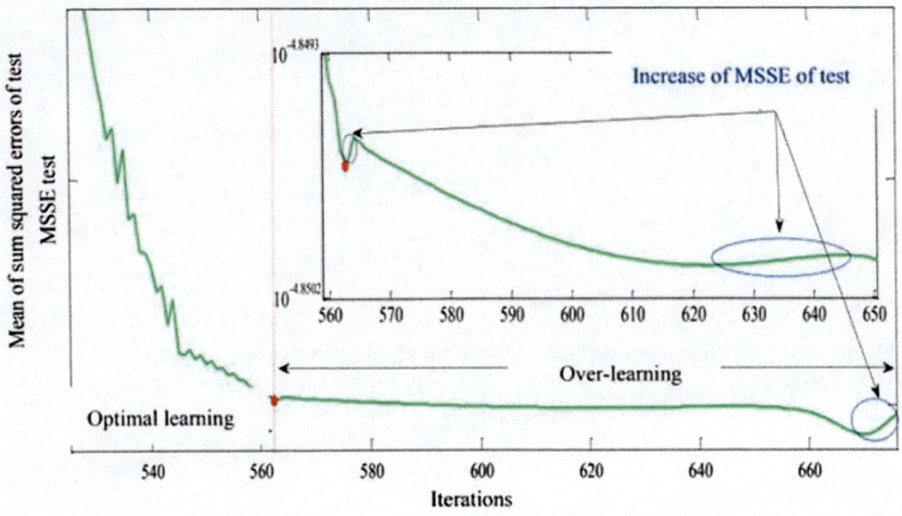

Fig. 6 Behavior of the network with seven electrical parameters during over-learning

5.3 *Evolution of the Seven Electrical Parameters According to Temperature and Irradiance*

In this section, we show through Figs. 7, 8, 9, 10, 11, 12 and 13 how the predicted values by our feedforward artificial neural network match very well with the target values of the R_s, R_{sh}, I_{ph}, I_{s1}, I_{s2}, n_1 and n_2. Otherwise, these figures illustrate how the temperature and/or the irradiance influence the value of each parameter.

The more the temperature increases, the more the value of the series resistance Rs increases also (Fig. 7). Indeed, the metallic contacts and connections which are located at the level of each cell, generate losses when the temperature rises to considerable values.

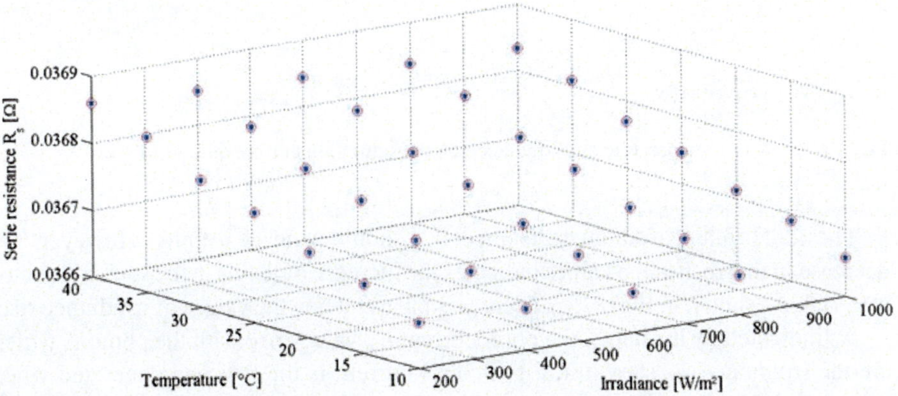

Fig. 7 Impact of temperature and irradiance on the series resistance R_s

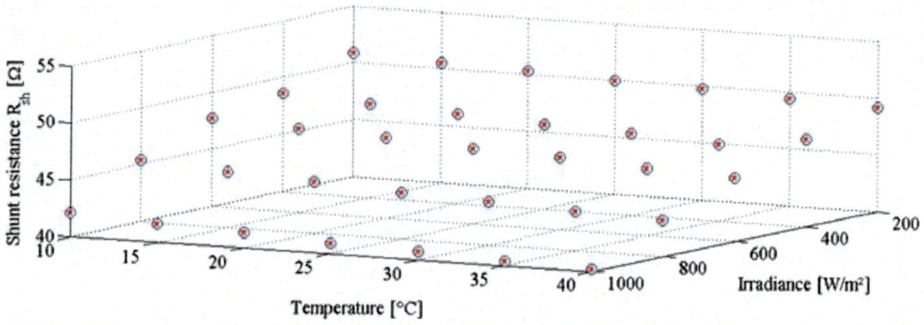

Fig. 8 Impact of temperature and irradiance on the shunt resistance R_{sh}

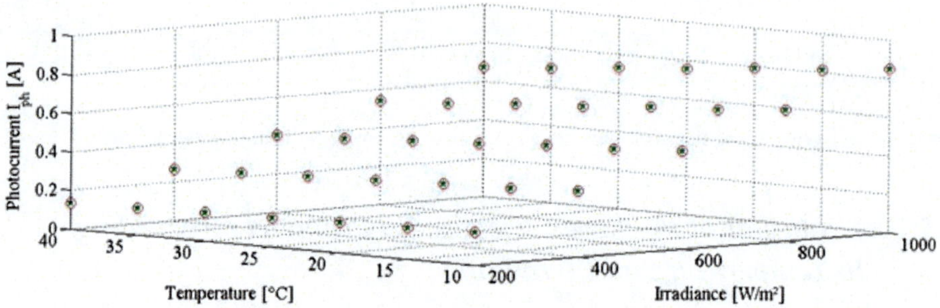

Fig. 9 Impact of temperature and irradiance on the photocurrent I_{ph}

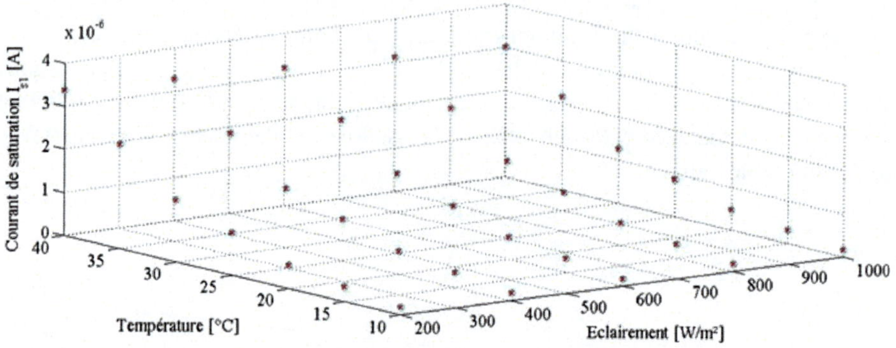

Fig. 10 Impact of temperature and irradiance on the saturation current of diode 1 I_{s1}

The ideal value of shunt resistance R_{sh} should tend to infinity. However, the decrease in this resistance values indicates the increase of losses in the semiconductor material. As shown in Fig. 8, R_{sh} decreases clearly when the values of irradiance rise.

As indicated by his name, the photocurrent I_{ph} values rise with the photons which are the irradiance, as shown in Fig. 9. This current is the first one generated when the solar cell is illuminated.

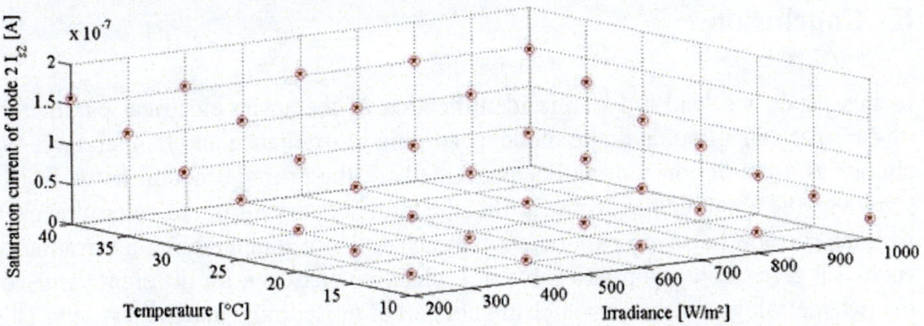

Fig. 11 Impact of temperature and irradiance on the saturation current of diode 2 I_{s2}

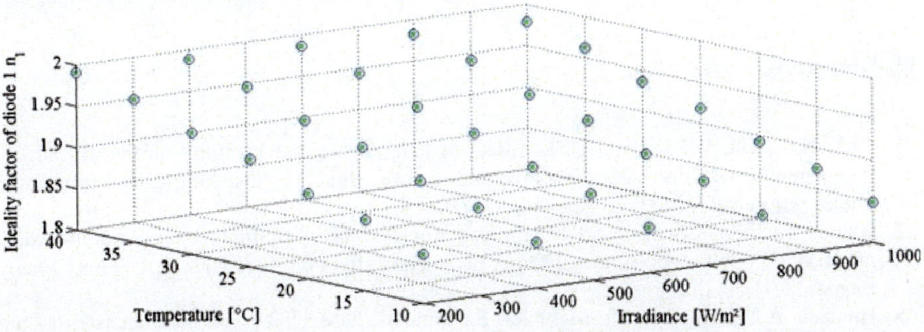

Fig. 12 Impact of temperature and irradiance on the ideality factor of diode 1 n_1

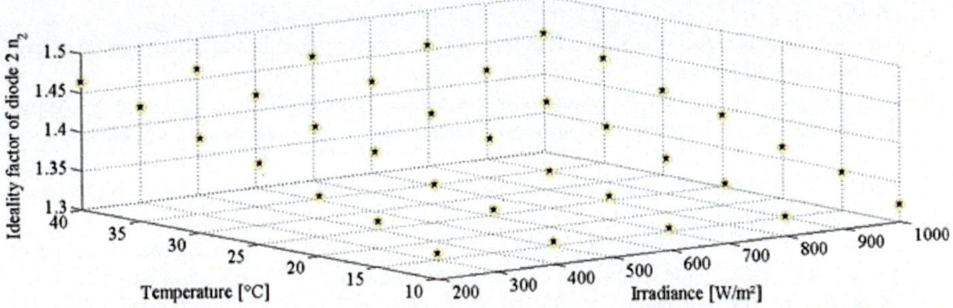

Fig. 13 Impact of temperature and irradiance on the ideality factor of diode 2 n_2

The two parameters Is1 and Is2 are more influenced by temperature. these two parameters present the saturation current of the two PN junctions. The latter model the intrinsic nature of the solar cell. in fact, these two currents depend to the movement of the electrons at the level of each junction, whose temperature is the main factor for the electron's agitation.

Likewise, for the ideality factor of the two PN junctions (n_1 and n_2), the values of these two parameters increase with increasing temperature.

6 Conclusion

In this paper, we lead a dynamic identification of the seven electrical parameters of the solar cell's double diode model according to irradiance and temperature. We choose as a prediction tool an adequate method: the artificial neural network. As shown by the obtained results the artificial neural network allows to determine each parameter from the seven parameters for each value of temperature and irradiance such as it gives the appropriate value of each parameter even for different values of the two metrological factors, which are not part of the learning and the test sets. This quality of generalization qualifies the neural network as an optimal method based on the black box principle.

References

1. Park, N.C., Oh, W.W., Kim, D.H.: Effect of temperature and humidity on the degradation rate of multicrystalline silicon photovoltaic module. Hindawi Publishing Corporation Int. J. Photoenergy, 1–9 (2013)
2. Ahmad, T., Sobhan, S., Nayan, F.: Comparative analysis between single diode and double diode model of PV Cell: concentrate different parameters effect on its efficiency. J. Power Energy Eng. **4**(3), 31–46 (2016)
3. Humada, A.M., Hojabri, M., Mekhilef, S., Hamada, H.M.: Solar cell parameters extraction based on single and double-diode models: a review. Renew. Sustain. Energy Rev. **56**, 494–509 (2016)
4. Barth, N., Jovanovic, R., Ahzi, S., Khaleel, M.A.: PV panel single and double diode models: Optimization of the parameters and temperature dependence. Solar Energy Mater. Solar Cell (2016)
5. Benghanem, S.M., Alamri, S.N.: Modeling of photovoltaic module and experimental determination of serial resistance. J. Taibah Univ. Sci. (2009)
6. Dreyfus, A.G., Martinez, J.-M., Samuelides, M.M., Gordon, B., Badran, F., Thiria, S.: Apprentissage Statistique: Réseaux de neurones, Cartes topologiques Machines à vecteurs supports. Eyrolles
7. Hornic, K.: Multilayer feedforward networks are universal approximators. Neural Netw. **2**, 359–366 (1989)
8. Rodrigues, E.M.G., Melício1, R. Mendes, V.M.F., Catalão, J.P.S.: Simulation of a solar cell considering single-diode equivalent circuit model. RE&PQJ **1**(9), 369–373 (2011)
9. Bensalem, S.: Effets de la température sur les paramètres caractéristiques des cellules solaires. Université Ferhat Abbas-Setif. (6 janvier 2011)
10. Petit. P.: Optimisation du transfert d'énergie dans les systèmes photovoltaïques. Thèse de doctorat, Université de Metz. (06 juillet 2011)
11. Blas, M.A., Torres, J.L., Prieto, E., Garcia, A.: Selecting a suitable model for characterizing photovoltaic devices. Renew. Energy **25**, 371–380 (2002)
12. Le cun, Y.: A theoretical framework for backpropagation, proceeding of the 1988 connetionist models summer school. 21–28 (1988)
13. Shahin, M.A., Maier, H.R., Jaksa, M.B.: Evolutionary data division methods for developing artificial neural network models in geotechnical engineering. University of Adelaide (November 2000)

An Assessment of Line Voltage Stability Indices to Select the Best Combination for Voltage Stability Prediction

Rabiaa Gadal, Faissal Elmariami, Aziz Oukennou, Naima Agouzoul, and Ali Tarraq

Abstract Power grids have undergone major structural changes in recent years and are constantly expanding as countries develop. This leads to complex phenomena and can cause network collapse. Therefore, there is an urgent need to continuously monitor and control electrical grids using voltage stability prevention methods, namely voltage stability indices. The objective of this work is to evaluate and compare different line indices considering their sensitivity to network disturbances. Choosing the best combination of the two most sensitive indices can be used to predict the point of voltage collapse. This study is carried out on the two standard bus test networks IEEE 5-bus and IEEE 14-bus. The results are validated using MATLAB software.

Keywords Blackout · FVSI · NCLVSI · NLSI · Lmn · LQP · Power flow · SFVSI · VSI · Voltage collapse · Voltage stability line indices

1 Introduction

The need for electrical energy is steadily rising over the world. As a result of this circumstance, the reinforcement of electrical networks has become an inevitable operation in order to maintain balance and meet customer demands. To reach this goal, additional power plants and transmission links must be built. However, technical, economic, and demographic limits make this operation difficult, if not impossible, for operators. As a result, the networks are reaching their maximum level of stability. This explains why blackouts are more common around the world. Dozens of blackouts have been registered in various countries over the last five years. The significant examples are presented in Table 1.

R. Gadal (✉) · F. Elmariami · N. Agouzoul · A. Tarraq
Energy and Electrical Systems Laboratory, Electrical Networks and Static Converters Team, National and High School of Electricity and Mechanics (ENSEM), Hassan II University of Casablanca, B.P 8118, Oasis, Casablanca, Morocco
e-mail: rabiaa.gadal@ensem.ac.ma

A. Oukennou
S.A.R.S. Team, National School of Applied Sciences (ENSA), UCA University, B.P 63, Safi, Morocco

M. Bendaoud et al. (eds.), *The Proceedings of the International Conference on Electrical Systems & Automation*, https://doi.org/10.1007/978-981-19-0035-8_8

Table 1 A brief summary of the significant blackouts took place since 2017: their consequences and causes

References	Blackout	Population affected (Million)	Blackout causes		
			Technique	Management	Climate
[3–6]	United States 2017	21	x	x	
	Usruguay 2017	3.4	x		x
	United States 2017	7.6	x	x	
	Sudan 2018	41.5	x	x	
	Azerbaïdjan 2018	8	x		
	Brazil 2018	10	x		
	Canada 2018	600 miles			x
	Venezuela 2019	30		x	
	Sri Lanka 2020	21	x		
	Pakistan 2021	200	x		

The analysis of the root causes was reported in [1] and led to grouping them into three categories: Management-related causes, technical causes, and causes related to climatic or other conditions. Furthermore, numerous phenomena are established between the cause and the blackout, and we note mostly the voltage stability. To deal with this predicament, researchers are focusing their efforts on developing methods for controlling and predicting this phenomenon. There are several approaches available, including curve elaboration (P-V, Q-V, and P-Q), modal analysis, continuation power flow method, and voltage stability indices. The latter is a practical tool for monitoring and determining the areas of the power system that requires compensation. According to [2], they are grouped into three categories: global, bus and line indices.

The line indices that are most frequently observed in the literature are given special consideration in this work. First, we will assess them to determine the sensitivity of each index and then suggest a mixed form that responds to the various network disruptions. All the evaluations will be carried out on two test networks: IEEE 5-bus and IEEE 14-bus.

The remainder of this paper is organized as follows: Section two introduces a set of line indices. The third portion focuses on assessing these indices. The new proposed index NCLVSI is presented in section four. The fifth part ends with a summary.

2 Overview of Some Line Voltage Stability Indices

Line voltage stability indices play an important role in the monitoring of the electrical network. They permit measuring the distance of the network operating point from the collapse point, and identifying the most critical line, they are simple, easy to

Table 2 Recap of some line voltage stability indices

Ref	Indicator/Formula	Nature of inputs	Number of inputs	Assumption	Critical value	Controlled power
[9]	$Lmn = 4\frac{QrX}{(Vs.\sin(\theta-\delta))^2}$	X, Q, V, Θ, δ	5	–	1	Q
[10]	$FVSI = 4\frac{QrZ^2}{Vs^2X}$	Z, Q, V, X	4	Y = 0, δ = 0	1	Q
[11, 12]	$SFVSI = 4\frac{Vr}{Vs}(1-\frac{Vr}{Vs})$	V	1	Y = 0, δ = 0	1	–
[13]	$LQP = 4\frac{X}{Vs^2}(Qr + Ps^2\frac{X}{Vs^2})$	X, P, Q, V	4	Y = 0, δ = 0	1	Q, P
[14]	$VSI = 4Qr\frac{R^2+X^2}{X.Vs^2}$	Q, R, X, V	4	Y = 0, δ = 0	1	Q
[15]	$NLSI = 4\frac{\Pr R+QrX}{Vs^2}$	P, R, Q, X, V	5	Y = 0, δ = 0	1	Q, P

where Vr, Vs: Voltage magnitude of (Sending bus, receiving bus), Qr: Reactive power in receiving bus, δ: Voltage angle, θ: Line angle, $Z = R + jX$: Line impedance,$FVSI$: Fast Voltage Stability Index,$SFVSI$: Simplified Fast Voltage Stability Index, *Lmn*: Line Stability Index, LQP: Line stability factor, VSI: Voltage Stability Indicator and $NLSI$: Novel Line Stability Index

implement and require a fairly short calculation time. Some of the available line indices existing in the literature are summarized in Table 2.

Line indices have been widely used in the literature. As there is also a combination proposal of these indices to obtain a new-found index. For example, in [7] a suggested index is based on the average of three indices FVSI, LPQ, Lmn to benefit from the advantages of each one. But this combination was not carried out regarding the sensitivity of those indices towards active and reactive loads. Another combination of the two indices Lmn and FVSI is proposed in [8], but this approach missed the evaluation of the resulting index in terms of active load variation.

3 Assessment of Some Line Voltage Stability Indices

3.1 Selection of Test Network

Our evaluation is performed around two standard networks: IEEE 5-bus and IEEE 14-bus as shown in Figs. 1 and 2. The data for both systems is taken from [16, 17].

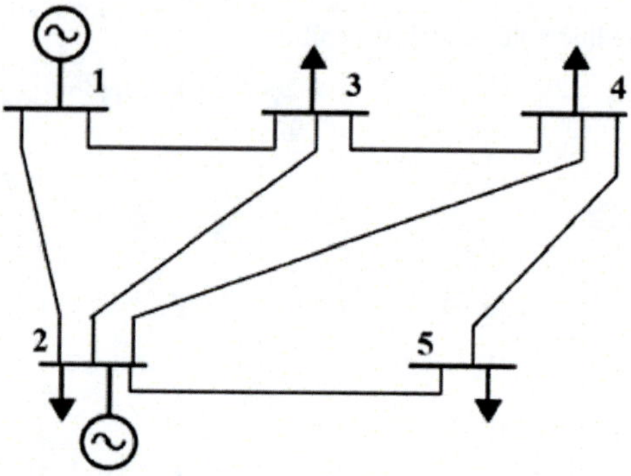

Fig. 1 Test network for IEEE 5-bus [16]

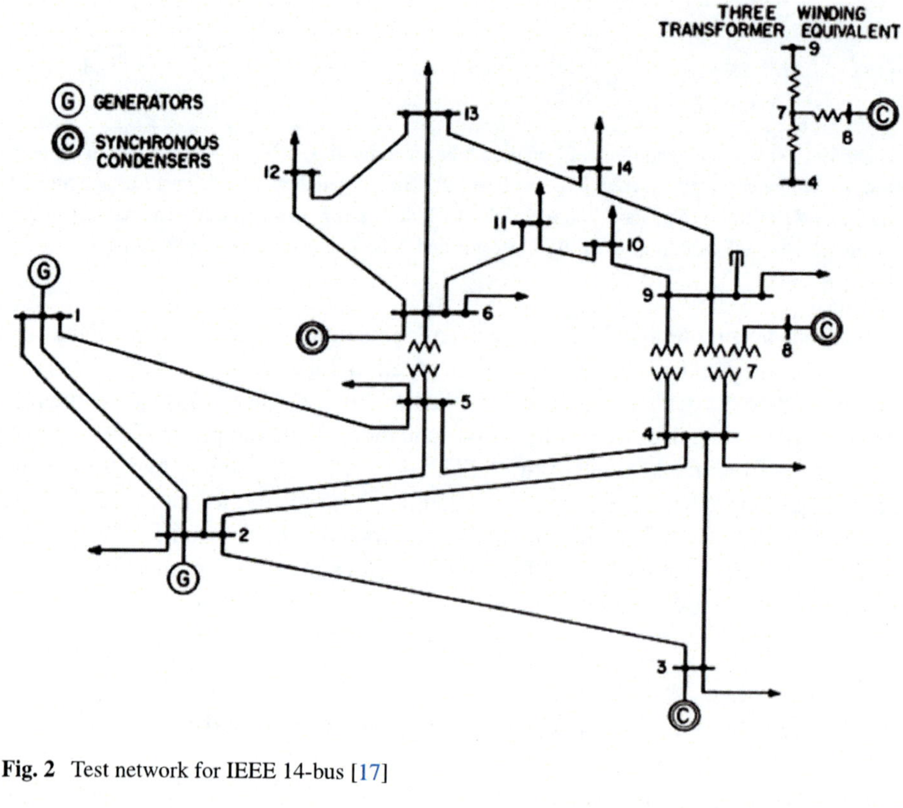

Fig. 2 Test network for IEEE 14-bus [17]

3.2 *Methodology for Evaluating Line Voltage Stability Indices*

The following process guides our examination: First, we insert the bus and line data into Matlab. Then we run the power flow. After the calculation of the line indices,

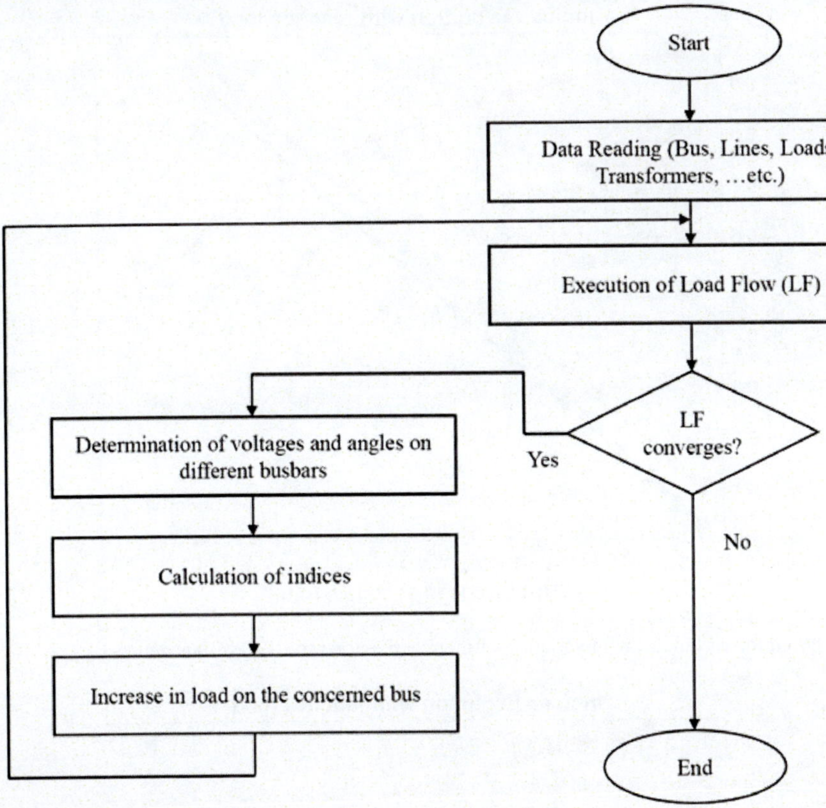

Fig. 3 Flowchart of the methodology

an increase of the reactive or active load is made. This procedure as shown in Fig. 3 is continued until the power flow ceases to converge (Fig. 3).

3.3 *Results and Simulations*

The evaluation is executed for the line (2–5) and the line (14–9) for the IEEE 5-bus and IEEE 14-bus network respectively. For these same lines, the indices reported in Table 2 are calculated and compared to each other according to the variation of the active and reactive load.

Reactive load variation:

Figures 4 and 5 illustrate the evolution of indices as a function of reactive load variation at two lines (2–5) and (14–9) for the IEEE 5 and IEEE 14-bus network respectively.

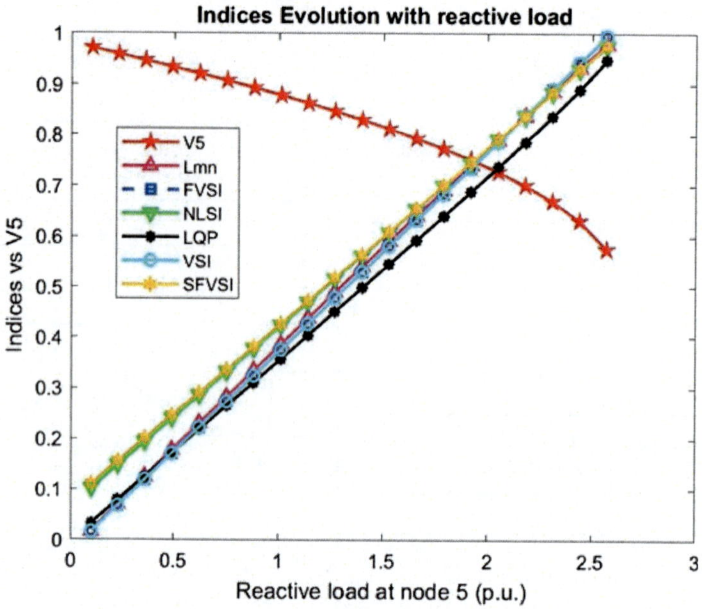

Fig. 4 Evolution of indices with variation of the reactive load for IEEE 5-bus network

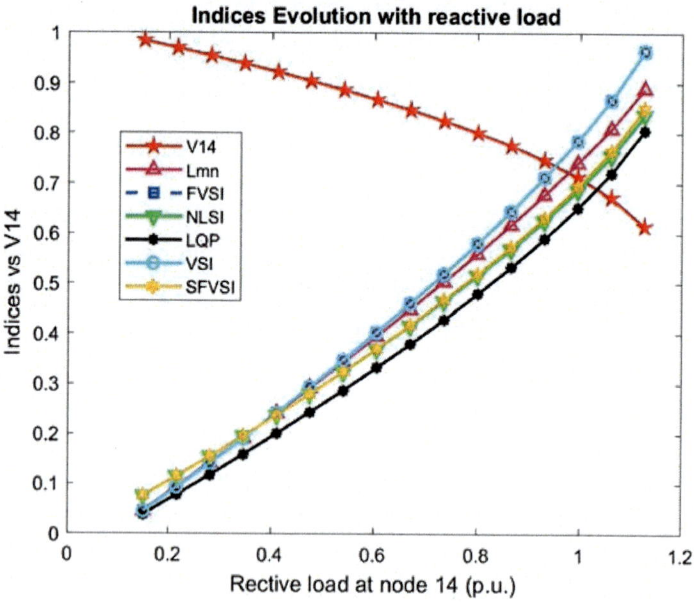

Fig. 5 Evolution of indices with variation of the reactive load for IEEE 14-bus network

As both Figs. 4 and 5 show, all indices undergo a significant variation according to the increase of the reactive load. Indeed, it is clear from the two figures that the Lmn and VSI indices have a high sensitivity towards the reactive load and take the most critical values. Also, based on Fig. 4 VSI seems to be the most sensitive index because it starts with a low value 0.04526 and reaches the critical value 0.965319 for a reactive load of 1.126 pu. Another comparison of VSI and LQP indices shows that the slope of VSI is higher than that of LQP. Therefore, VSI has proven its superiority for reactive load monitoring.

Active load variation:

The simulations' results of indices evolution as a function of active load variations, at two lines (2–5) and (14–9) for the IEEE 5 and 14-bus networks respectively, are shown in the following two Figs. 6 and 7.

The analysis of Figs. 6 and 7 allows to notice that VSI and Lmn indices have an insignificant variation range which is very far from the critical value, as a result, they have an insignificant sensitivity referred to the active load variation. However, according to the 5-bus network, the SFVSI index has a wide range of variation; it starts with a low value of 0.1105 and reaches a significant value of 0.93402 pu for a maximum active load of 3.981 pu. Also, it should be noted that despite the significant variation range of LQP compared to SFVSI as shown in Fig. 6, the LQP index shows a nonlinear evolution and a slope lower than that of SFVSI. Therefore, we deduce that SFVSI has a strong sensitivity towards the evolution of the active load.

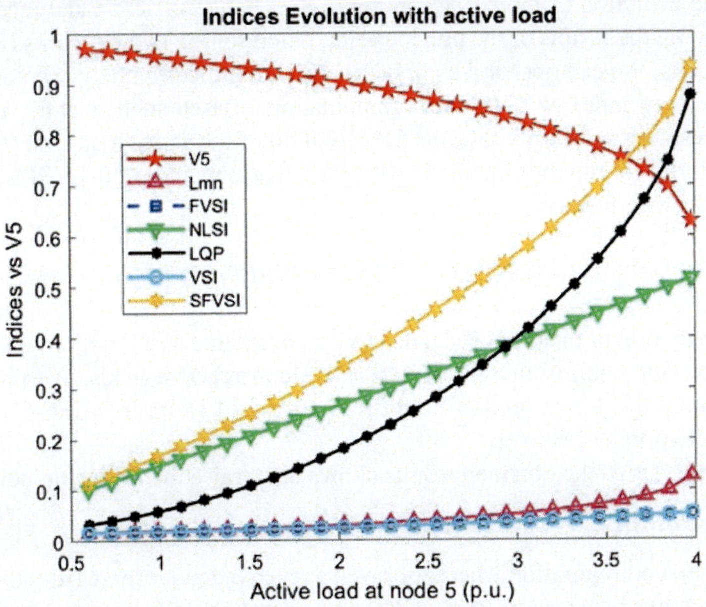

Fig. 6 Indices evolution with variation of active load for IEEE 5-bus network

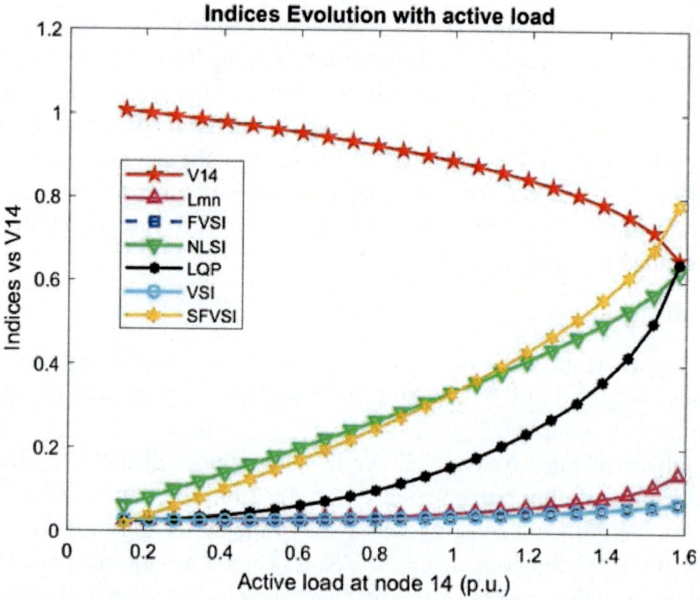

Fig. 7 Indices evolution with variation of active load for IEEE 14-bus network

4 Novel Combined Line Voltage Stability Index (NCLVSI)

Given the evolution of indices according to the change in active and reactive load, and based on the results of the previous simulation for the two networks IEEE 5 and IEEE 14 bus. We can opt for a combination of two indices namely: SFVSI and VSI to obtain a new index NCLVSI. This combination is based on the fact that the SFVSI showed a strong sensitivity towards the evolution of the active load, and VSI proved its superiority for monitoring the reactive load. The new index NCLVSI is presented in the following form:

$$\text{NCLVSI} = \text{Max}(\text{SFVSI}, \text{VSI}) \tag{1}$$

The evolution of the proposed index will be evaluated as a function of the change in voltage and stability margin of two active and reactive loads. We represent in Figs. 8 and 9 the power profiles established for 5 and 14 bus of IEEE 5- and IEEE 14-bus networks respectively.

The analysis of the obtained results allows us to raise the following points:

In the case of IEEE 5-bus network:

- In the first configuration where we saved a reactive load reserve (rqq) of 2.533 pu and an active load reserve (rpp) of 3.386 pu, we noticed a low value of the NCLVSI index of 0.1100, which is logical due to the fact that the maximum limit of the two active and reactive loads (Maxp, Maxq) is 2.633 and 3.986 pu successively,

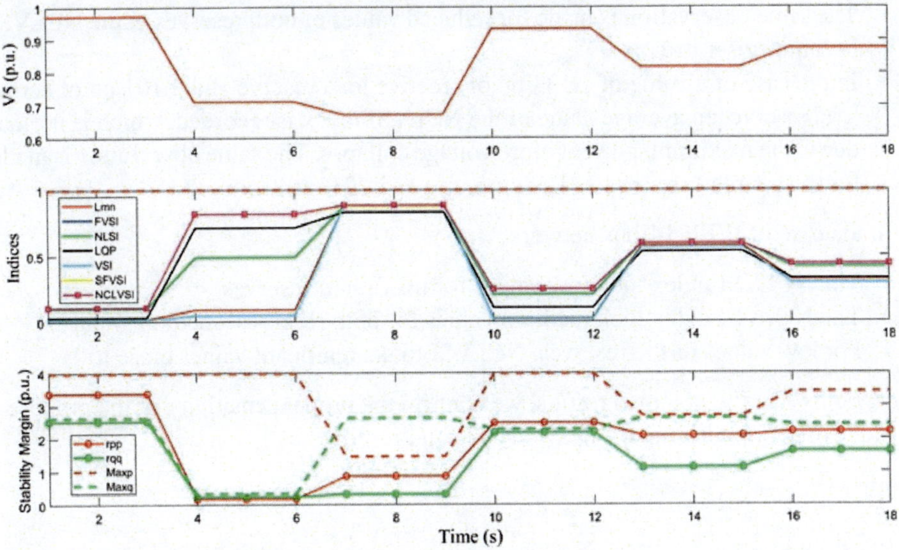

Fig. 8 NCLVSI evaCn under load variation for IEEE 5- bus

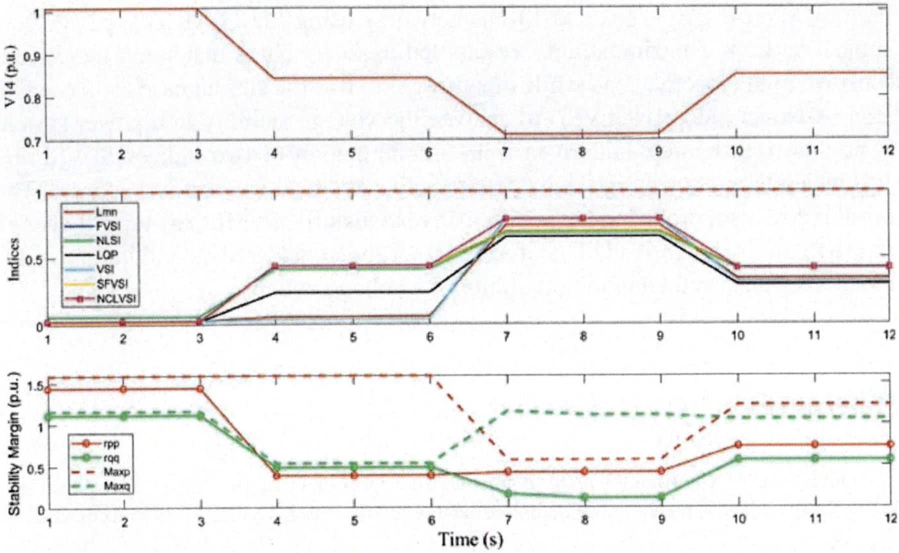

Fig. 9 NCLVSI evaluation under load variation for IEEE 14-.bus

- After a significant reduction of the active load reserve, the index reached a significant value of 0.8190,
- The index re-reached these low values when the reserves rpp and rqq approached their maximum values,

- The same observation is made for reduced values of both reserves while NCLVSI became closer to 1,
- In the case of a value of 1.133 pu of reactive load reserve and 2.101 pu of active load reserve, an average value of the NCLVSI index is recorded, which is normal due to the maximum power before voltage collapse. The same observation is made for the couple case (rqq = 1.624 pu, rpp = 2.221 pu).

In the case of IEEE 14-bus network:

- The NCLVSI index took low values for maximum reserves,
- The NCLVSI index took medium values for both reserves' medium values,
- For low values of the reserves, NCLVSI took significant values close to 1.

Following the obtained results, we confirm the proper sensitivity of the proposed index in the monitoring of the two active and reactive loads.

5 Conclusion

This study gave a comparative assessment of voltage stability line indices. It was performed on IEEE 5- bus and 14-bus networks using MATLAB as a power flow simulation tool. The simulations' results led us to conclude that some indices are sensitive to the reactive load while others are not. It is the similar case for the active load. A Novel index (NCLVSI) to analyze the voltage stability in a power system is proposed, this index is derived from a combination of two indices SFVSI and VSI which have shown high sensitivity to active and reactive load respectively. The novel index, also, proved to be more sensitive to disturbances for any type of change in active and reactive load. Therefore, it could be considered as an alternative for monitoring and predicting the proximity of a voltage collapse.

References

1. Oukennou, A.: Contribution au développement des indicateurs de stabilité de la tension et leur utilisation dans la recherche de l'emplacement optimal des FACTS dans les réseaux électriques (2021)
2. Oukennou, A., Sandali, A.: Voltage stability indices sensitivity evaluation under load variation in electrical power system. In: Proceedings 2017 International Conference on Electrical and Information Technologies, vol. 2018-Janua, pp. 1–5 (2018)
3. Alhelou, H.H., Hamedani-Golshan, M.E., Njenda, T.C., Siano, P.: A survey on power system blackout and cascading events: research motivations and challenges. Energies **12**(4):682, Feb (2019). https://www.mdpi.com/1996-1073/12/4/682/htm
4. Pozzebon, S., Britton, B.: Venezuela power outage leaves most of country in dark—CNN. https://edition.cnn.com/2019/03/08/americas/venezuela-blackout-power-intl/index.html (2019)

5. Farzan, Z.: Island-wide Power Cut due to an issue in the transmission system, https://www.newsfirst.lk/2020/08/17/island-wide-power-cut-due-to-an-issue-in-the-transmission-system/ (2020)
6. Masood, S.: Much of Pakistan loses power in massive blackout—The New York times, https://www.nytimes.com/2021/01/09/world/asia/pakistan-blackout-power-failure.html (2021)
7. Dazahra, M.N., Elmariami, F., Belfqih, A., Boukherouaa, J.: Optimal location of SVC using particle swarm optimization and voltage stability indexes. Int. J. Electr. Comput. Eng. **6**(6), 2581–2588 (2016). https://doi.org/10.11591/ijece.v6i6.11921
8. Samuel, I.A., Katende, J., Awosope, C.O.A., Awelewa, A.A.: Prediction of voltage collapse in electrical power system networks using a new voltage stability index. Int. J. Appl. Eng. Res. **12**(2), 190–199 (2017)
9. Moghavvemi, M., Omar, F.M.: Technique for contingency monitoring and voltage collapse prediction. IEE Proc. Gener. Transm. Distrib. **145**(6), 634–640 (1998)
10. Musirin, I., Rahman, T.K.A.: Estimating maximum loadability for weak bus identification using FVSI. IEEE Power Eng. Rev. **22**(11), 50–52 (2002)
11. Sahari, S., Abidin, A.F., Rahman, T.K.A.: Development of artificial neural network for voltage stability monitoring. In: Proceedings National Power Engineering Conference, pp. 37–42 (2003)
12. Abdul Rahman, T.K., Jasmon, G.B.: New technique for voltage stability analysis in a power system and improved loadflow algorithm for distribution network. In: Proceedings 1995 International Conference on Energy Management and Power Delivery EMPD, vol. 2. pp. 714–719 (1995)
13. Mohamed, S.Y.A., Jasmon, G.B.: A static voltage collapse indicator using line stability factors. J. Ind. Technol. **7**(1), 73–85 (1989)
14. Chattopadhyay, T.K., Banerjee, S., Chanda, C. K.: Impact of distributed generator on voltage stability analysis of distribution networks under critical loading conditions. In: 2014 1st International Conference of Non Conventional Energy Search Clean Safe Energy, ICONCE Iconce, pp. 288–291 (2014)
15. Yazdanpanah-Goharrizi, A., Asghari, R.: A novel line stability index (NLSI) for voltage stability assessment of power systems. In: The 7th WSEAS International Conference on Power Systems, pp. 164–167 (2007)
16. Roy, R.B., Alahakoon, S., Arachchillage, S.J.: Grid impacts of uncoordinated fast charging of electric ferry (2021)
17. Boudreaux, J.A.: Design, simulation, and construction of an IEEE 14-Bus power system (2018)

An Intelligent Control of a Variable Speed Wind Turbine Based on DFIG for Maximum Power Capture

Aicha Bouzem, Othmane Bendaou, and Bousselham Samoudi

Abstract Due to the advancement of wind turbine industry technologies, the variable-speed wind turbine (WT) coupled with a doubly fed induction generator (DFIG) has attracted considerable interest due to its several potential advantages over other wind turbine concepts. To contribute to this fast-growing development, different wind systems control strategies are looking to become more intelligent to operate the WT around its optimum operation with high security and reliability. Practically, the WT's efficiency can be achieved by extracting the maximum possible amount of power from the wind. In this context, we are particularly interested in this work to implement an intelligent control of a variable speed wind turbine based on a DFIG using the intelligent artificial techniques, by combining an artificial neural network Maximum Power Point Tracking (ANN-MPPT) and intelligent Indirect vector control by stator field alignment (ANN-IFOC). The ANN-MPPT strategy aims to extract a maximum of power from the wind to operate the WT around its optimum operation independently of the system parameters, the aerodynamic characteristics, and the wind speed measurement. While the intelligent IFOC uses ANN-controllers to optimize the generator's active and reactive powers. The efficiency of the presented control system topology is confirmed by the simulation results acquired using Matlab/Simulink software; the obtained results are satisfactory and confirm the ability of the suggested approach to maintaining the system operating at the desired response.

Keywords Wind energy · Doubly fed induction generator DFIG · Maximum power point tracking (MPPT) · Indirect field oriented control (IFOC) · Artificial neural network (ANN)

A. Bouzem (✉) · O. Bendaou
Department of Physics, Faculty of Sciences, Abdelmalek Essaadi University, Sebta Avenue, 93002 Tetouan, Morocco
e-mail: aicha.bouzem@etu.uae.ac.ma

B. Samoudi
Civil Engineering, Energetic and Environment Department, National School of Applied Sciences, Abdelmalek Essaadi University, Sidi Bouafif Ajdir, 32003 Al Hoceima, Morocco
e-mail: b.samoudi@uae.ac.ma

M. Bendaoud et al. (eds.), *The Proceedings of the International Conference on Electrical Systems & Automation*, https://doi.org/10.1007/978-981-19-0035-8_9

1 Introduction

The Wind Energy Conversion Systems play an important role as an alternative solution for energy generation due to many motivations, such as the absence of greenhouse gas pollution, the unlimited availability of driving sources, and the absence of hazardous waste [1].

Under trends of using wind energy sources, different control strategies for wind power systems are seeking to become more efficient and more intelligent in order to meet the future electricity demands and the huge distribution of it. The WT's efficiency can be increased by capturing the maximum available power from the wind and operating the WT at its optimum operation under rapidly varying environmental conditions, by implementing advanced control strategies [1, 2].

In this context, many Maximum Power Point Tracking (MPPT) approaches have been proposed and applied, such as Hill Climb Search (HCS), perturbation and observation (P&O), incremental conductance (IncCond) [2, 3]. In recent years, MPPT seeks to be more powerful in order to overcome the many limitations of traditional MPPT, such as the inaccurate wind speed measurement, the degradation of the aerodynamic properties of the aeroturbine with time, and the variation of the climatic properties from one site to another.

Artificial intelligence techniques have demonstrated new solutions in industrial processes due to their many benefits compared to conventional computational systems, and especially, they have become a perfect solution for highly sensitive control mechanisms and non-linear models, due to their ability to provide highly accurate and faster responses [4, 5].

The current work presented in this paper intends to implement a proposed Maximum Power Point Tracking (MPPT) approach based on artificial intelligent techniques (Artificial Neural Network (ANN)) accompanied by an intelligent Indirect field oriented control (ANNIFOC) to optimize the energy generated from a DFIG coupled with a variable speed wind turbine and connected to the grid.

The ANN-MPPT strategy aims to capture maximum power under varying wind speeds, independently of the system parameters, the aerodynamic characteristics, and the wind speed measurement. While the intelligent IFOC aims to control the generator's active and reactive powers and avoid any disruption caused by characteristic uncertainty, which can affect the quality of the supplied energy.

This work is presented as follows: Sect. 2 describes the model of our wind turbine energy system and the conventional MPPT strategy. The suggested ANN-MPPT is explained in Sect. 3. Section 4 includes the mathematical model of DFIG in the d-q reference frame and the ANN-IFOC strategy. While Sect. 5 is reserved for the presentation of simulation results obtained using Matlab/Simulink, to prove the effectiveness of the proposed control strategy to ensure the system's optimal operation. Finally, some conclusions are summarized in Sect. 6.

2 Modeling of the Wind Turbine and the MPPT Strategy

2.1 Modeling of the Wind Turbine

The mechanical power extracted from the wind can be represented as follows [6]:

$$P_{aero} = \frac{1}{2} C_P(\lambda, \beta)\, \rho\, S\, v^3 \tag{1}$$

where, ρ is the air density, S is the turbine swept area, and v denotes the wind speed (m/s).

The power coefficient $C_P(\lambda, \beta)$ (Fig. 1) measures the turbine's aerodynamic efficiency; which is affected by the size of the blade, the angle of the blade's orientation (β) and the speed ratio (λ) [7]:

$$C_p(\lambda, \beta) = 0.5176\left(\frac{116}{\lambda_i} - 0.4\beta - 5\right)\exp\left(\frac{-21}{\lambda_i}\right) + 0.0068\lambda \tag{2}$$

where:

$$\lambda_i = \frac{1}{\lambda + 0.08\beta} - \frac{0.035}{\beta^3 + 1} \tag{3}$$

$$\lambda = \frac{\Omega_t R}{v} \tag{4}$$

The wind exerts a mechanical torque on the turbine shaft, which can be expressed by the following equation:

$$C_t = \frac{P_t}{\Omega_t} = \frac{1}{2}\rho\,\pi\, R^3 V^2\, C_c(\lambda, \beta) \tag{5}$$

where: $C_c = \frac{C_p}{\lambda}$; $\Omega_t = \frac{\Omega_m}{G}$; $C_g = \frac{C_t}{G}$.

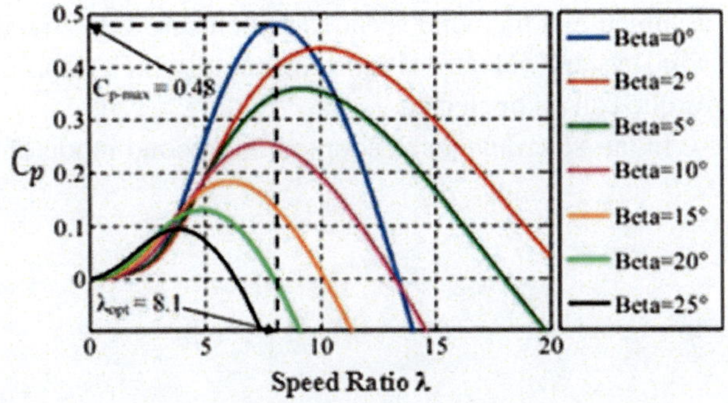

Fig. 1 Power coefficient $C_P(\lambda, \beta)$

The turbine and mechanical rated speeds are denoted respectively by Ω_t and Ω_m, while G denotes the Gearbox ratio.

2.2 The Maximum Power Point Tracking Strategy (MPPT)

The optimum operation of the WT is achieved by running the turbine at the maximum aerodynamic power coefficient ($C_{p,max}$), which corresponds to an optimum value of λ and (λ_{opt}, β_{opt}), while λ is adjusted to its optimal value by controlling indirectly the rotor speed.

According to (2), (3), and Fig. 1, the desired value of $C_{p\_max} = 0.48$ of our system is reached for $\lambda_{opt} = 8.1$, and $\beta = 0$.

$$\Omega_{\mathrm{m,opt}} = \frac{\mathrm{G\,v}\,\lambda_{opt}}{\mathrm{R}} \tag{6}$$

By combining (4), (5), and (6) we obtain the expression of $\mathbf{C_{em\_ref}}$ and $\boldsymbol{P_{aero\_max}}$:

$$C_{em\_ref} = \frac{\pi\,\rho\,R^5 \mathrm{C_{p,max}}}{2\,G^3\,\lambda_{opt}{}^3}\,\Omega^2_{\mathrm{m,opt}} \tag{7}$$

$$P_{aero\_max} = \frac{\pi}{2}\,\rho\,R^2 \left(\frac{R\,\Omega_{\mathrm{m,opt}}}{G\,\lambda_{opt}}\right)^2 \mathrm{C}_{p\_max} \tag{8}$$

Generally, we can distinguish two main modes of MPPT [2–8]:

With Mechanical Velocity Regulation. This method requires wind speed measurement using an anemometer or an array of anemometers and using PI regulators, which increases the cost of the system. On the other hand, in practice, an accurate wind speed measurement is difficult to reach, and an imprecise measurement necessarily reduces the system's reliability and degrades the extracted power. For these reasons, most of the WT systems are currently controlled without mechanical velocity regulation.

Without Mechanical Velocity Regulation. For this second control structure, it is assumed that the wind speed variations are very low in steady state compared to the wind system's electrical time constants, which implies that the turbine's acceleration torque can be neglected.

In our study, we have adopted the second mode of MPPT (Fig. 2).

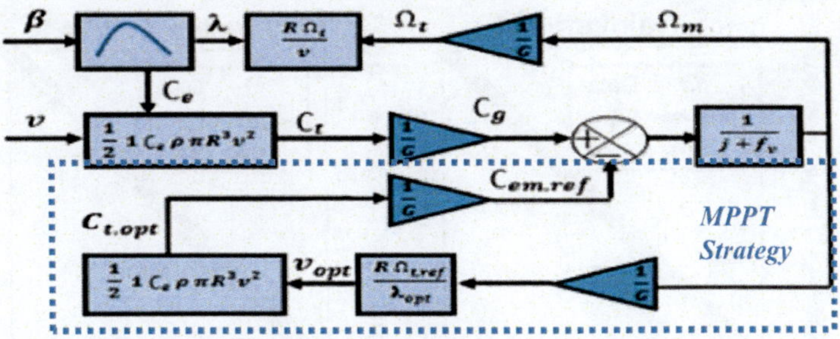

Fig. 2 Block diagram of the wind turbine system with MPPT

3 Maximum Power Extraction by Artificial Neural Networks ANN-MPPT

The ANN algorithms are inspired by the biological processes of the human brain. His goal is to accomplish specified activities or functions based on a collection of connected artificial neurons. The NNs take in data and train themselves to be able to anticipate the outputs of a new set of similar types of data.

The following equations represent the expression of the output yk of a neuron k [9]:

$$v_k = \sum_{x=1}^{x=n} (W_{ki} \cdot x_i + b) \tag{9}$$

$$y_k = f(v_k) \tag{10}$$

The ANNs are constructed from a number of layers (input layer, output layer, and hidden layers). The neurons in the layers are connected by channels, and each of these channels is assigned a value called a weight (W_{ki}), which is adjusted during the training phase using a learning algorithm. The inputs (x_i) are multiplied to the corresponding weights and summed with the bias (b) to generate output after passing through a threshold function called the activation function (f).

In our case, the training set (the input and target), used for forming the ANN for MPPT [8], is obtained from the conventional MPPT provided in Sect. 2 (Fig. 2) using Matlab/Simulink. The net is implemented to determine the optimal electromagnetic torque that requires the turbine to operate at its optimum operation whatever the wind speed.

Figure 3 shows the regression curves for the ANN, which provide highly significant information about the performance of the ANN training, validation, and testing, based on the value of R and the distribution of the data along the adjustment line (Fit). For our ANN, we show that R is equal to 1, which indicates that the ANN has successfully trained up to 100%, and we also observe that all data points are

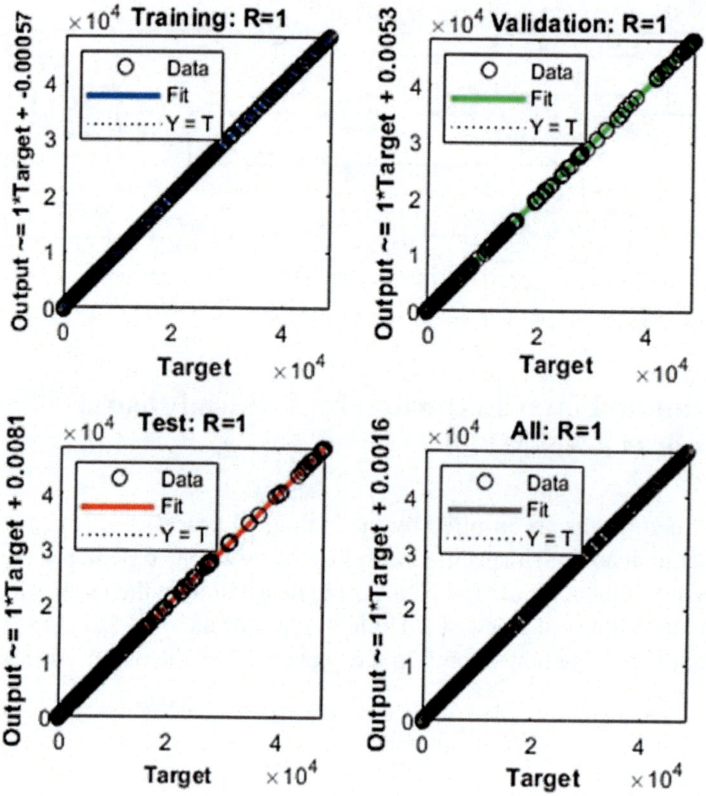

Fig. 3 Regression curves for ANN-MPPT

aligned with the Fit line, confirming that the ANN has precisely comprehended the relationship between the input and output data.

The ANN controller's performance is assured by selecting the optimum number of neurons and hidden layers through a series of tests. In our case, we reached the best architecture by using one hidden layer containing 10 neurons.

4 Modeling and Intelligent Field Oriented Control of the DFIG

The configuration of DFIG can be presented in d-q reference by the following equations [10]:

- Stator and rotor voltages

$$\begin{cases} V_{sd} = R_s I_{sd} + \frac{d}{dt}\varnothing_{sd} - \omega_s \varnothing_{sd} \\ V_{sq} = R_s I_{sq} + \frac{d}{dt}\varnothing_{sq} + \omega_s \varnothing_{sq} \\ V_{rd} = R_r I_{rd} + \frac{d}{dt}\varnothing_{rd} - (\omega_s - \omega_r)\varnothing_{rd} \\ V_{rq} = R_r I_{rq} + \frac{d}{dt}\varnothing_{rq} + (\omega_s - \omega_r)\varnothing_{rq} \end{cases} \quad (11)$$

- Stator and rotor flux

$$\begin{cases} \varnothing_{sd} = L_s I_{sd} + L_m I_{rd} \\ \varnothing_{sq} = L_s I_{sq} + L_m I_{rq} \\ \varnothing_{rd} = L_r I_{rd} + L_m I_{sd} \\ \varnothing_{rq} = L_r I_{rq} + L_m I_{sq} \end{cases} \quad (12)$$

- Electromagnetic torque

$$T_{em} = \frac{L_m p}{L_s}\left(\varnothing_{sd} I_{rq} + \varnothing_{sq} I_{rd}\right) \quad (13)$$

- The active and reactive stator powers

$$\begin{cases} P_s = V_{sd} I_{sd} + V_{sq} I_{sq} \\ Q_s = V_{sq} I_{sd} - V_{sd} I_{sq} \end{cases} \quad (14)$$

To control the electrical power generated by DFIG, we will control the exchange of the active and reactive power between the DFIG stator and the grid, based on indirect vector control, by aligning the stator flux with the d-axis (Fig. 4), accounting for the coupling terms, and compensating for them with a two-loop system [4].

We have: $\varnothing_{sd} = \varnothing_s$, $\varnothing_{sq} = 0$, and $\frac{d}{dt}\varnothing_{sd} = 0$.

The voltage equations of DFIG can be simplified as:

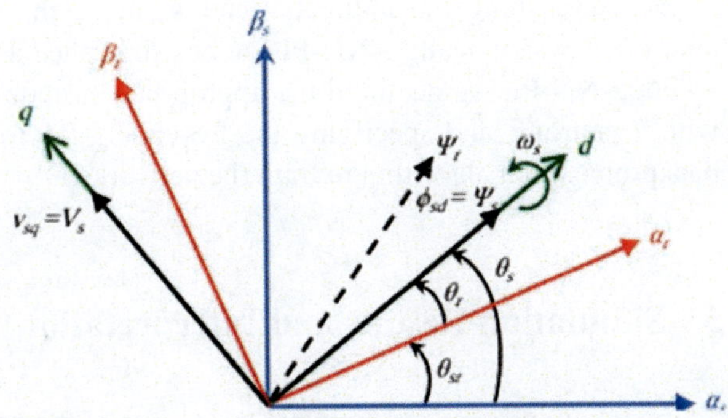

Fig. 4 Orientation of the stator flux of the DFIG

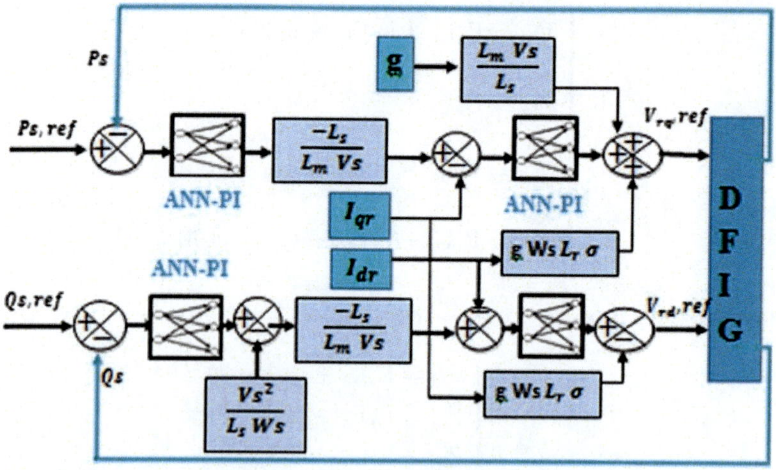

Fig. 5 Global block diagram of ANN-indirect field-oriented control technique

$$\begin{cases} V_{sd} = 0 \\ V_{sq} = \mathrm{Vs} = \omega_s \varnothing_s \\ V_{rd} = R_r I_{rd} + Lr\,\sigma \dfrac{d}{dt} I_{rd} - g\omega_s\,\sigma\, I_{rq} \\ V_{rq} = R_r I_{rq} + Lr\,\sigma \dfrac{d}{dt} I_{rq} + g\omega_s\,\sigma\, I_{rd} + g\dfrac{L_m Vs}{L_s} \end{cases} \tag{15}$$

With: $\sigma = 1 - \frac{L_m^2}{L_s L_r}$.

By replacing (12) and (15) in (14) the generator's active and reactive powers can express by:

$$\begin{cases} P_s = -\dfrac{L_m Vs}{L_s} I_{rq} \\ Q_s = -\dfrac{L_m Vs}{L_s} I_{rd} + \dfrac{L_m \varnothing_s}{L_s} \end{cases} \tag{16}$$

By establishing the indirect vector strategy, the global block diagram of the controlled system using ANN-PI can be established as shown in Fig. 5 [4].

For ANN-PI, we acquired the appropriate structure by taking one hidden layer with 7 neurons, and specifying the Levenberg–Marquardt (LM) algorithm as the backpropagation algorithm to train the networks.

5 Simulation Results and Interpretation

To model our system and simulate the results of the control presented in this paper, we used MATLAB/Simulink software (Fig. 6). The system operates in a closed loop,

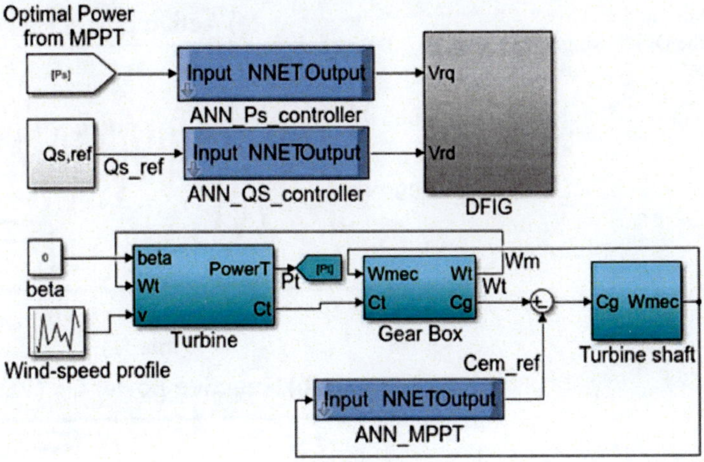

Fig. 6 Global block diagram of the proposed control

where the ANN-MPPT control block provides the reference power to the IFOC loop for a DFIG of 10 kW.

The results shown in Fig. 7 are related to the ANN-MPPT strategy, which shows that the proposed MPPT required the system to maintain the power coefficient around its maximum value $C_{pmax} = 0.48$ (Fig. 7b). Moreover, the extracted mechanical power

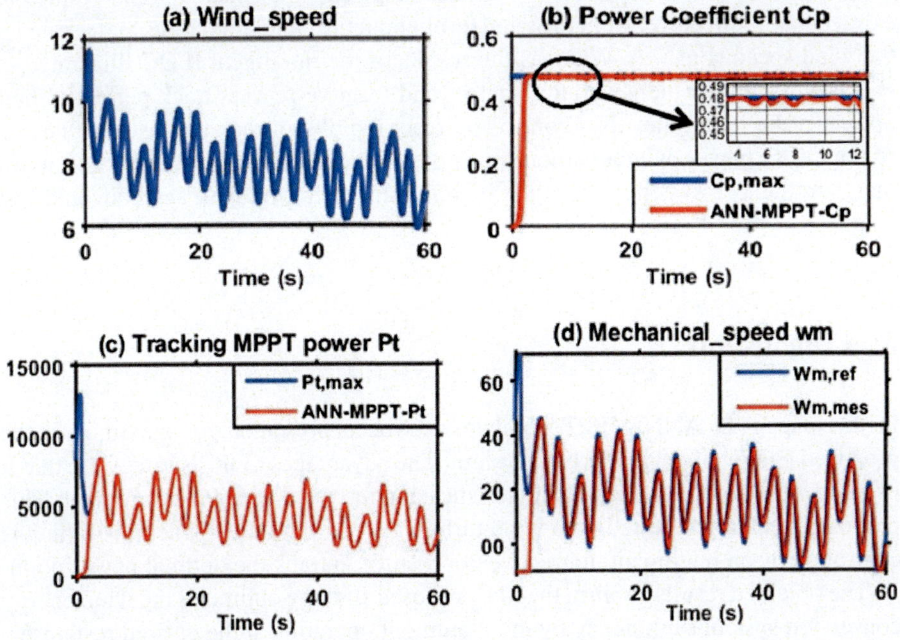

Fig. 7 Simulation results of the ANN-MPPT

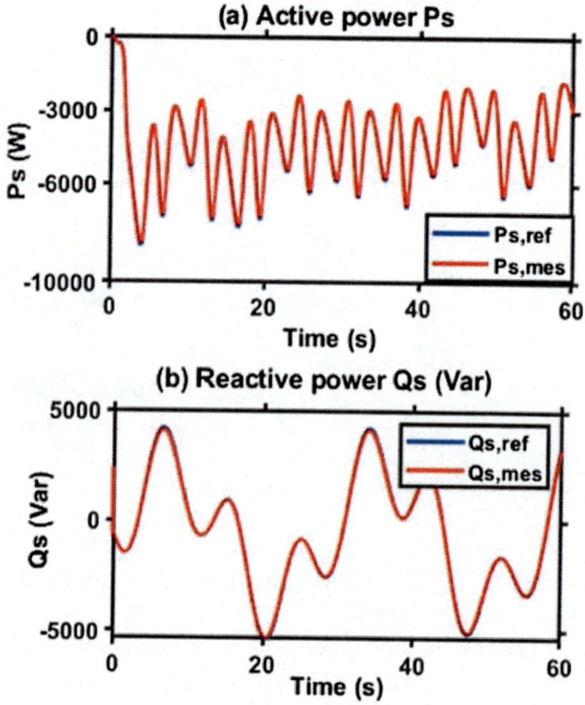

Fig. 8 Active and reactive power of the DFIG using ANN-IFOC

(Fig. 7c) and the turbine rotational speeds (Fig. 7d) track their optimal values to extract maximum power from the wind throughout the simulation time, and whatever the wind speed (Fig. 7a). Whereas the results of the intelligent IFOC illustrated in Fig. 8 indicate that the generator's active and reactive powers track perfectly their optimal values, and the ANN controllers react rapidly to track the fast fluctuations of the DFIG power, which demonstrates the efficiency of the suggested control to maintain the WT system working at its optimum operation with accuracy and fast response.

6 Conclusions

In this paper, an ANN MPPT regulator has been presented for maximizing the produced energy of a wind turbine system. The advantages of this strategy are that it is independent of the system characteristics and the wind speed measurement. This proposed MPPT was tested on a wind turbine based on a DFIG generator, which is controlled by an intelligent indirect vector control to track the optimal power point.

The obtained results confirm that the proposed strategy optimizes the wind energy conversion system's efficiency by maintaining it operating at the desired response.

References

1. Kumar, Y., Ringenberg, J., Depuru, S.S., Devabhaktuni, V.K., Lee, J.W., Nikolaidis, E., Andersen, B., Afjeh, A.: Wind energy: trends and enabling technologies. Renew. Sustain. Energy Rev. **53**, 209–224 (2016). ISSN 1364-0321. https://doi.org/10.1016/j.rser.2015.07.200
2. Mousa, H.H.H., Youssef, A.-R., Mohamed, E.E.M.: State of the art perturb and observe MPPT algorithms based wind energy conversion systems: a technology review. Int. J. Electr. Power Energy Syst. **126**(Part A), 106598 (2021). ISSN 0142-0615. https://doi.org/10.1016/j.ijepes.2020.106598
3. Musunuri, S., Ginn, H.L.: Comprehensive review of wind energy maximum power extraction algorithms. In: IEEE Power and Energy Society General Meeting, pp. 1–8 (2011). https://doi.org/10.1109/PES.2011.6039023
4. Djeriri, Y., Meroufel, A., Allam, M.: Artificial neural network-based robust tracking control for doubly fed induction generator used in wind energy conversion systems. J. Adv. Res. Sci. Technol. (2015). ISSN: 2352-9989
5. Aamir, M.: On replacing PID controller with ANN controller for DC motor position control. Int. J. Res. Stud. Comput. **2**, 21–29 (2013). https://doi.org/10.5861/ijrsc.2013.236
6. Tahir, K., Belfedal, C., Allaoui, T., Doumi, M.: Proposal of a new hybrid control strategy for dynamic performance improvement of wound field synchronous generator-based wind turbines. J. Renew. Sustain. Energy **7**, 043113 (2015). https://doi.org/10.1063/1.4926768
7. Djeriri, Y.: Lyapunov-based robust power controllers for a doubly fed induction generator. IJEEE **16**(4), 551–558 (2020). https://doi.org/10.22068/IJEEE.16.4.551
8. Samir, L., Said, G., Mustapha, D., Youcef, S.: A neural MPPT approach for a wind turbine. In: 6th International Conference on Systems and Control (ICSC), pp. 210–214. IEEE, Batna, Algeria (2017). https://doi.org/10.1109/ICoSC.2017.7958689
9. Nour Ali, M.: Improved design of artificial neural network for MPPT of grid-connected PV systems. In: 2018 Twentieth International Middle East Power Systems Conference (MEPCON), pp. 97–102. IEEE, Cairo, Egypt (2018). https://doi.org/10.1109/MEPCON.2018.8635202
10. Allam, M., Dehiba, B., Abid, M., Djeriri, Y.: Etude comparative entre la commande vectorielle directe et indirecte de la Machine Asynchrone à Double Alimentation (MADA) dédiée à une application éolienne 13 (2014)

Potential and Challenges in Small Hydro Power Projects in India

Alok Bora, Saurabh Awasthi, and Nafees Ahamed

Abstract The whole world is focusing and striving for clean, green, and renewable energy sources to meet the ever growing energy demand and reduce the environmental impact due to excessive use of fossil fuels. Every country has to contribute in reducing its over-dependence on non-renewable energy and switching to greener options to stop and reverse the damages of ecological imbalance with urgency and sincerity. India being the second most populous country and seventh largest in the world in terms of the land area plays an important role in this. The Indian Himalayan Region (IHR) is a hilly region blessed with many rivers and natural ascent making it ideal for hydropower generation. This paper will discuss small hydropower potential and challenges in India and give probable solutions on ways to overcome these challenges in order to utilize the maximum small hydro potential of the country. Also, it will be discussed how small hydro (including mini, micro and pico) power can play a major role in the generation of clean energy and reduce India's over-dependence on fossil fuels and in turn reducing her carbon footprint. The findings and results will be helpful for other countries as well which have small hydro potential available.

Keywords Small hydropower · Micro hydropower · Pico hydropower · Renewable energy

1 Introduction

With an ever increasing population and never-ending demand for electricity, the journey of India from a developing country to a developed superpower is constantly getting delayed. To accomplish this, one of the most important things—if not the most important—the country will have to achieve is energy sufficiency or in other words state of energy surplus. The reason this is of paramount importance is that energy lies at the very foundation of all other industry or infrastructure projects. Add to this the automobile industry, which is on the very cusp of technological change to go from

---

A. Bora (✉) · S. Awasthi · N. Ahamed
Department of Electrical and Electronics and Communication Engineering, DIT University, Dehradun, Uttarakhand 248009, India

M. Bendaoud et al. (eds.), *The Proceedings of the International Conference on Electrical Systems & Automation*, https://doi.org/10.1007/978-981-19-0035-8_10

petrol/diesel to electric in the coming years. This in combination with a multitude of other factors is sure to cause an exponential rise in demand and consumption of electrical energy not only in India but all over the world. As a result, to meet this steep surge of energy demand, India will have to up its energy production, relying mainly on renewable energy sources along with a robust and extensive transmission and distribution system if it has to avoid the impending energy crisis and continue on the growth trajectory. Energy consumption is directly proportional to the quality of life of people in any country or specific region and is a universally accepted yardstick to measure the same.

The over-dependence of our country on imported crude oil causes depletion of foreign reserves, which could be used in a better way elsewhere. Pollution due to the burning of fossil fuels and subsequent environmental imbalance is another major concern associated with it. The only solution to this problem lies in developing clean and renewable energy sources within the country which will go a long way in putting India on the fast track of growth and development. For this to actually happen, each state will have to pitch in by identifying and utilizing the renewable energy resources it has been bestowed with, to the maximum. Almost every Indian state has at least one or more combinations of renewable resources available comprising solar, hydro, wind, tidal, etc. Some of these technologies may not be economically viable at the moment but the ones with promise must be pursued aggressively. Although a lot of development has taken place in the country in the last decade in the space of renewable energy, under the Ministry of New and Renewable Energy (MNRE), experts believe that we have barely scratched the surface. A huge amount of untapped renewable energy is waiting to be harnessed. This paper will focus on tapping the small hydropower (SHP) potential in India while also discussing the challenges in its implementation and probable solutions.

There is no doubt that being a renewable energy source hydropower will play a crucial role in contributing to energy generation in the future. International Energy Association (IEA) has stated that hydropower will continue to become a major energy source among the various renewable energy sources in near future [1]. Hydropower is widely considered as a leading renewable energy source and is gaining importance in energy generation all over the world [2–4]. Currently, India ranks 5th in hydropower generation after China, Brazil, United States of America and Canada [5]. Utilizing small hydro potential to the maximum along with medium and large hydro potential is the need of the hour and will go a long way in making India energy surplus.

2 Methodology

The methodology of this paper comprises of acquiring, comparing, and analyzing data from various sources such as scientific literature presented in journals and conferences, various reputed websites including websites of various ministries of Government of India for authentic information on hydropower projects especially small micro and pico hydropower plants. Also, visit to SHP plants to gather first-hand information by Engineers running such plants about the challenges being faced in this sector. An extensive literature review is carried out on the said topic and on the basis of collected data, possible solutions for specific problems in the SHP sector are suggested which can be used as a reference for future SHP works by investors and other interested people in this sector.

3 Classification of Hydro Power Plants

Hydro power plants are classified into large medium and small depending on their capacity. Different countries have different parameters to categorize them. In India their classification is done as follows:

1. Large hydro—>100.00 MW
2. Medium hydro—>25.00 to ≤100.00 MW
3. Small hydro*—>2.00 to ≤25.00 MW
   - Mini hydro—>100.00 kW to ≤2.00 MW
   - Micro hydro—>5.00 to ≤100.00 kW
   - Pico hydro—≤5.00 kW.

*In general any hydropower project less than 25.00 MW is broadly classified as small hydro, therefore, mini, micro and pico are sub-categories of small hydro.

3.1 Formula of Power Generated from Hydro

The formula for hydropower generated is given as

$$P = \eta \rho Q g h \tag{1}$$

where,

- P power generated (W)
- η dimensionless efficiency of the turbine (approx. 0.9)
- ρ density of water (1000 kg/m^3)
- Q volumetric flow rate of water (m^3/s)
- g acceleration due to gravity (9.8 m/s^2)
- h height difference between inlet and outlet (m).

The above equation clearly shows that three out of the five parameters are constants (η, ρ, g), in the given framework, which means that the power developed is directly proportional to 'Q' and 'h' i.e. volumetric flow rate of water and height difference between inlet and outlet.

3.2 Advantages of SHP

As SHP is a renewable energy source, it has the obvious benefits of being clean, sustainable, and has zero fuel cost compared to non-renewable energy sources like fossil fuels. In addition to these benefits, it has several other advantages also, which makes it among the best and most lucrative in non-renewable category.

The additional advantages of SHP are as follows [6]:

(1) High efficiency (70–90%) which is by far the best among all energy technologies.
(2) High Capacity Factor (typically 40–50%). Capacity factor is defined as ratio of actual energy produced by an energy generating unit in a given time period, to the hypothetical maximum possible (i.e. energy produced from continuous operation at full rated power).
(3) High level of predictability, varying with annual rainfall patterns. Compared to solar, and wind, flow and volume of rivers can be more accurately predicted.
(4) Slow rate of change i.e. the output power varies only gradually from day to day and not from minute to minute as in the case of solar and wind.
(5) It is a long lasting and robust technology. SHP systems can be readily engineered to last 50 years and more.
(6) Can go from stopped condition to full power in just a few minutes.
(7) Installation and commissioning for micro and pico power plants can be done in a few weeks as compared to years in the case of large and medium hydro power plants.
(8) In case of natural calamity like flood etc., the major components can be dissembled and later assembled easily to protect them from getting damaged.
(9) Since SHPs like micro and pico power plants are generally close to load centers, therefore, extensive transmission and distribution network is not necessary in such cases leading to savings in the amount of copper wire, poles, step-up and step-down transformers, and other equipment used. Also, transmission and distribution losses are reduced to a minimum since power does not have to be transmitted over long distances.

4 India's Energy Scenario and SHP Potential

In the past, India has always been an energy deficient country. Although energy generation has always shown year-on-year growth and the percentage of deficiency has decreased over time but the state of energy surplus has eluded the country until now. The bulk of energy generation has been done by fossil fuels especially coal due to its easy availability and low cost. But the last decade has shown tremendous growth in the renewable energy sector and the deficiency has been reduced to an all-time low.

4.1 India's Energy Scenario of Last 12 Years

Before we discuss SHP potential in Indian Himalayan Region (IHR), we have to look at the overall picture of energy generation in the country to understand why it is of upmost importance. Table 1 shows the Indian energy scenario from 2009–10 to 2020–21 [7].

With India being the second-most populous country in the world, its energy requirement is also huge. Figure 1 shows the difference in energy requirement and availability in India from 2009–10 to 2020–21 and Fig. 2 shows the difference in peak demand and peak met in the same period.

Table 1 shows that the energy generation in the country has substantially increased from 2009–10 to 2020–21. The energy deficit has been brought down from double digits to decimal point, which is a commendable achievement. Also, the energy production of the country is on the verge of becoming fully surplus in the coming few years. The total installed capacity of India (as on 28/02/2021) as per CEA is 379,130 MW or 379.13 GW. The figure shows that in the past two Fiscal Years (FY), the energy deficit has been around 0.5% and 0.4% respectively. Although it is a small percentage and work done to bring it down to this level is commendable, we also have to look at the dark side of this data. In a country with a population of almost 1.4 billion, this deficiency means that there are still many villages with a sizable population that does not have access to electricity even at this age. Moreover, there are millions of other people in the country who do not have the luxury of

Table 1 Indian energy scenario from 2009–10 to 2020–21

Year	Energy				Peak			
	Requirement	Availability	Surplus (+)/deficit (−)		Peak demand	Peak met	Surplus (+)/deficit (−)	
	(MU)	(MU)	(MU)	(%)	(MW)	(MW)	(MW)	(%)
2009–10	830,594	746,644	−83,950	−10.1	119,166	104,009	−15,157	−12.7
2010–11	861,591	788,355	−73,236	−8.5	122,287	110,256	−12,031	−9.8
2011–12	937,199	857,886	−79,313	−8.5	130,006	116,191	−13,815	−10.6
2012–13	995,557	908,652	−86,905	−8.7	135,453	123,294	−12,159	−9.0
2013–14	1,002,257	959,829	−42,428	−4.2	135,918	129,815	−6103	−4.5
2014–15	1,068,923	1,030,785	−38,138	−3.6	148,166	141,160	−7006	−4.7
2015–16	1,114,408	1,090,850	−23,558	−2.1	153,366	148,463	−4903	−3.2
2016–17	1,142,929	1,135,334	−7595	−0.7	159,542	156,934	−2608	−1.6
2017–18	1,213,326	1,204,697	−8629	−0.7	164,066	160,752	−3314	−2.0
2018–19	1,274,595	1,267,526	−7070	−0.6	177,022	175,528	−1494	−0.8
2019–20	1,291,010	1,284,444	−6566	−0.5	183,804	182,533	−1271	−0.7
2020–21[a]	1,155,130	1,150,891	−4239	−0.4	190,198	189,395	−802	−0.4

[a]Upto February 2021 (Provisional)

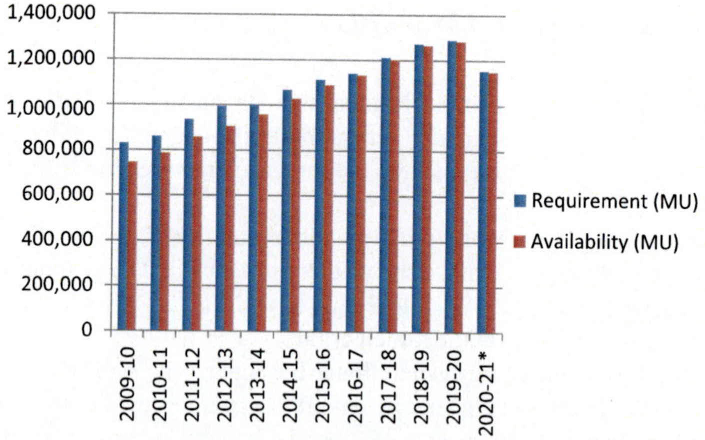

Fig. 1 Shows the difference in energy requirement and availability in India from 2009–10 to 2020–21

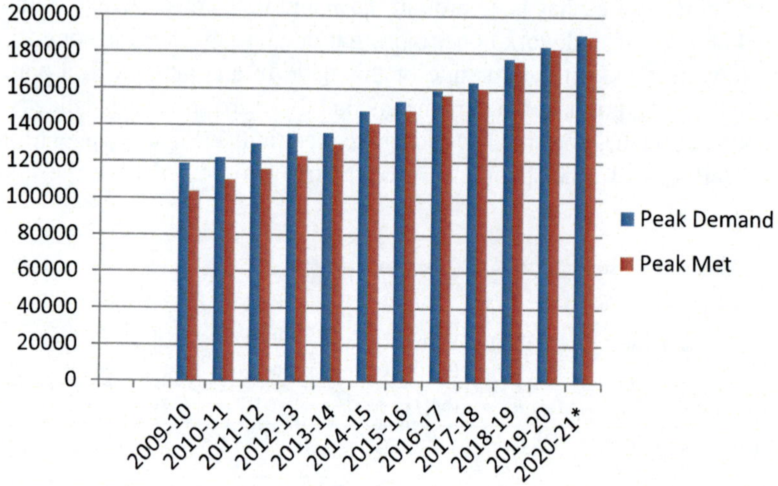

Fig. 2 Shows the difference in peak demand and peak met India from 2009–10 to 2020–21

uninterrupted electricity supply, with constant load-shedding part of their already hard life. This load-shedding sometimes extends up to 12 hours every day at some places. Needless to say, this energy shortage is causing a major hindrance in many an individual's life and also the nation as a whole.

Going by the recent data (early 2021), the energy production currently is fluctuating between being deficit marginally for the most part and going a modest surplus occasionally. This modest energy surplus theory could be attributed to the fact that the country has gone under multiple lockdowns due to the COVID-19 pandemic from March 2020 up to mid-2021, resulting in the closure of various industries, educational institutions, commercial spaces, etc. This closure has caused an obvious reduction in demand and hence the surplus energy state may go back to deficit once the situation normalizes and demand goes back to its peak. Then there is also the perennial problem of uneven energy distribution among cities and villages and various states. For example, the North-Eastern states of India face a higher energy deficit than the national average. This may be due to a multitude of factors but the main reason would surely be transmission and distribution challenges due to the geographic location of these states. In fact, it would not be wrong to say that the bigger problem plaguing the country currently is not energy production but transmission and distribution. The last mile delivery of electricity to every home is quite a challenge in a country as geographically vast and diverse as India. With the current urgency and resolve of the power sector, it is only a matter of time that this will be fully accomplished. However, things need to move at a faster rate by ramping up energy generation especially by renewable methods. This is due to the fact that a delay of each day is causing wastage in the tune of hundreds and thousands of MWs of energy in the form of solar, hydro, wind, etc. The sooner and more of these renewable energies are tapped, the better it will be in the overall scheme of things. It is predicted by experts that by 2030, energy consumption will double and the demand will be tripled.

4.2 *India's Current Installed Capacity*

The energy generation in the country is done through four major categories comprising of thermal, large hydro, nuclear, and renewable energy sources. Table 2 shows the installed capacity of India as of 28/02/2021 [7].

Table 2 clearly shows that more than 60% of total generation is still dependent on thermal energy sources with coal alone contributing 53% of total generation. Coal although cheap is the biggest environmental polluter and so its usage has to be drastically reduced and substituted with green energy options. This will also be in

Table 2 Installed capacity of India as on 28/02/2021

S. No.	Fuel	MW	% of total
1.	Total thermal	233,171	61.5
a.	Coal	201,085	53.0
b.	Lignite	6620	1.7
c.	Gas	24,957	6.6
d.	Diesel	510	0.1
2.	Hydro (large)	46,209	12.2
3.	Nuclear	6780	1.8
4.	RES (MNRE)[a]	91,154	24.5
	Total	379,130	100

[a]RES include SHP, biomass gasifier, biomass power, urban and industrial waste power, solar and wind energy

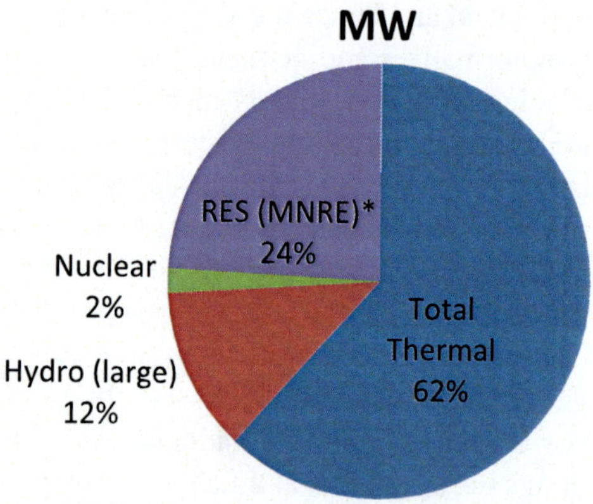

Fig. 3 Shows the share of various fuels in total energy production in India currently (upto February 2021)

line with India's energy goals to boost its renewable power capacity to 175 GW by 2022 (Fig. 3).

4.3 *SHP Potential of India*

India is blessed with many rivers due to which the potential of hydropower is quite substantial in the country. Table 3 shows SHP data of the top five states/UT [8].

The estimated potential of SHP in India is 21,133.65 MW from 7133 sites located in different states of the country [8]. The cumulative capacity target of 5000 MW

Table 3 SHP data of top-five state/UT as per annual report of MNRE, 2020–21

S. No.	State/UT	Total potential		Total projects installed (upto 2020–21)		Projects under implementation		% utilization of energy potential (including projects under implementation)
		Nos	Total capacity	Nos	Total capacity	Nos	Total capacity	
1.	Karnataka	618	3726.49	170	1280.73	3	13.00	34.71
2.	Himachal Pradesh	1049	3460.34	196	911.51	13	151.60	30.72
3.	Arunachal Pradesh	800	2064.92	156	131.11	9	6.05	6.64
4.	Uttarakhand	442	1664.31	102	214.32	14	28.58	14.59
5.	UT of Jammu and Kashmir	103	1311.79	19	146.34	6	31.90	13.58
	Total of all states and UTs	7133	21,133.62	1134	4750.46	96	450.80	24.61

for SHP by 2022 is well within reach as an aggregate capacity of 4750.46 MW has been achieved by 31st December 2020 through 1134 SHP projects. In addition, 96 projects of aggregate capacity of 450.80 MW are at various stages of implementation. The overall target is to achieve grid-connected renewable energy power projects of 175,000 MW or 175 GW by 2022. Four out of the top five in the list comprise of states/Union Territories (UT) in IHR facing more or less common challenges in terms of SHP implementation. The IHR region consists of 12 states/UTs comprising of UT of Jammu and Kashmir and Laddakh and states of Himachal Pradesh, Uttarakhand, Sikkim, Arunachal Pradesh, Nagaland, Manipur, Tripura, Meghalaya, and Assam. The IHR consists of many rivers and hilly regions which makes it ideal for hydropower production. The actual utilization of SHP in India and particularly IHR is still very low. To substantially increase this utilization percentage, the various stake holders in SHP production will have to work together and remove the bottlenecks which cause hindrance in the implementation of this clean and green technology.

4.4 SHP Potential in Indian Himalayan Region (IHR)

As discussed earlier, the hydropower developed depends mainly on volumetric flow rate of water and height difference between inlet and outlet. IHR is abundantly blessed with both these parameters in abundance. The region has dozens of rivers that are ice-fed and flow all year round (high 'Q'). Also, due to its geographic location, the region boasts of natural hilly terrain and a steep accent (high 'h').

5 Challenges and Risks in SHP

With every energy source, there are certain benefits and risks associated. As of now, there is no perfect or ideal energy source that can be practically harnessed to provide clean, sustainable, and cheap electricity for years and decades to come on a worldwide scale. Although continuous research and development is being done in this sector by the top scientists of the whole world, we are still a few years away, if not decades, to arrive at a possible energy solution. As said earlier, every energy source whether renewable or non-renewable has pros and cons. SHP too has its fair share of downsides. It is important to know the challenges and risks associated with SHP and ways to minimize them before starting any project in this sector.

5.1 Identification of Risks

Any feature, be it benefit or risk can be subdivided into either tangible (quantitative) or intangible (qualitative) features. Tangible features are those which can be expressed in costs and benefits i.e. in monetary terms. Conversely, intangible features are those which cannot be readily valued in money, for example, environmental and socio-economic risks.

The risks in hydropower projects, in general, can be broadly classified into seven categories and then further subdivided into various risk factors. It is important to mention that this is not a complete list but a selection of possible risks. The importance and emphasis of every kind of risk depend on the potential site, target group, technology, and the stage for an implementation of a hydropower plant.

5.2 *Classification of Risks in SHP*

The major classes of risk in small hydropower projects are as follows [9]:

(1) Technical Risk: This risk comprises technical related aspects in the project such as machinery, breakdown, operation and maintenance, as well as the delay from suppliers.
(2) Construction Risk: Construction of a dam is a huge capital investment project. It includes the construction schedule and construction budget risk.
(3) Financial Risk: This risk plays a major role in the small hydropower project. Financial resources, tax rate, and inflation risk come under this risk type. Financial resources can be subdivided into fund blockage and interest rate risk. The exchange rate risk comes into play only when foreign investment is made, which in the case of IHR, is negligible.
(4) Legal Risk: Obtaining the Legal clearances is one of the starting points in SHP projects. It consists of getting clearances as per regulatory. Approvals have to be taken at various levels and from different departments before actual implementation can begin. Public private partnership and norms and rule and regulation changes come under legal risks.
(5) Business Risk: This risk is associated with the situation in which the investor(s) in SHP project run into financial difficulties and not able to generate profit or live up to market expectations. The risk parameters under this category consist of capital cost, electricity price, generation, and modeling techniques.
(6) Environmental Risk: Environmental risks are those factors that may affect the environment or climate of the particular region. These include changes in river flow, precipitation, and flora and fauna.
(7) Socio-Economic Risk: This risk directly or indirectly affects the general population of the region where the SHP project is set up. This risk needs to be minimized to increase the safety and well-being of the local community. The various risks under this category are relocation and rehabilitation, loss of employment, affecting tourist places, noise pollution, changes in water quality, and soil erosion.

5.3 Additional Risks in IHR

In addition to above-mentioned risks, there are several additional risks in SHP which are present in IHR states of India. They are as follows:

(a) Risk of natural hazards:

- Being located in an active seismic zone region, the region is prone to high-intensity earth quakes (especially the state of Uttarakhand). To overcome this problem, a detailed study is required to identify the sites which are low risk and ideal for SHP.
- Floods are also a major concern that can damage not just small and medium but large hydropower plants as well. Floods can cause due to multiple reasons. Major reasons of flood include heavy rainfall, cloud bursts and breaking of glacier or its part.

(b) Damage to the electro-mechanical components of the SHP plant especially the turbine due to silt: Silt is the fine particles of sand that are present in the river. This silt continuously comes in contact with the blades of the turbine and corrodes it gradually over time. This results in a frequent change of turbine and other components which escalates the cost of the plant.

(c) Inaccessibility of project location due to lack of proper motorable roads increases the cost and delays the project substantially. This is because all the heavy equipment has to be transported on foot to the site which is often in a remote location. This may take weeks or even months. This not only delays the project but also puts the equipment at risk during transportation.

(d) Lack of skilled man power that is willing to work in remote locations.

(e) Due to the above reasons and high-risk factors, the interest rate on loans as well as insurance premiums on such projects are on the higher side. This financial roadblock comes across as a major deterrent among the private investors of SHP plants in the IHR states of India.

5.4 Major Natural Disasters in IHR in Recent Times

In the last decade, 2 major disasters have struck IHR region especially the state of Uttarakhand, and 4 more in the bordering countries of Nepal and China which also come in Himalayan region [10]. The chronology of these disasters is as follows:

(1) **Kedarnath disaster, Uttarakhand, India, 2013**

On 16th June, 2013, a flash flood of unseen magnitude hit Kedarnath, Uttarakhand, India. This flash flood occurred due to a cloudburst which caused the collapse of the bank (moraine wall) of the Chorabari glacier lake in Kedarnath [10]. This caused a chain reaction of landslide and huge debris flow causing widespread destruction in Uttarakhand. There was a huge loss of life and property as more than 5000 people lost their lives.

(2) **Chamoli floods, Uttarakhand, India, 2021**

On 7th February 2021, a huge chunk of rock (with snow and ice) detached from Rontigad Mountain and fell into Rontigad River [10–12]. This rock was about 550 m wide and had an estimated volume of 25 million m^3 and fell from an elevation of about 5600 m. This apparently caused huge mass of landslide debris to propagate along the steep slope downstream and pushing Rontigad and Rishiganaga rivers towards Dhauliganga which resulted in hyper-concentration and partly granular flow. The water level at downstream reaches (e.g. at Alakhnanda) exceeded the extreme level which evinces the signature of flash flood. The exact cause for this disaster is still under investigation but based on available information and analysis, the most convincing conclusion is that it was a landslide of a huge rock mass with thick ice/snow and their pulverization leading to debris flow-induced flash flood.

5.5 Effect of Chamoli Disaster on Hydro Power Projects

Due to the Chamoli disaster, Rishiganga hydropower project (13.2 MW, operational) in Rishiganga River near Raini village and Tapovan Vishnugad hydropower project (520 MW, under construction in Dhauliganga River) were severely damaged. Also, some other projects, namely Vishnuprayag HPP (400 MW, under operation) and Vishnugad Pipalkoti HPP (444 MW, under construction) were affected (e.g. closed) by the disaster.

Due to this disaster, the 520 MW Tapovan-Vishnugad hydel project has suffered an estimated loss of ₹1500 crore and it is unlikely that it will meet its scheduled commissioning target in 2023. Also, huge amount of silt is deposited in the project site which will take a considerable amount of time to be desilt by the authorities. The Geological Survey of India has found 13 of 486 glacial lakes in Uttarakhand to be vulnerable through remote sensing and multispectral data [10].

5.6 List of Disasters in Neighboring Countries of Himalayan Region

Apart from the above two major disasters in Uttarakhand, India, a few other major disasters occurred in the neighboring countries of Nepal and China in the last decade [10]. All these disaster hit territories share geographical similarities and therefore offer insights into damage done and precautionary measures that should be taken to minimize the losses.

1. Seti River disaster, Nepal, 2012
2. Jure landslide, Nepal, 2014
3. Glacial Lake Outburst Flood (GLOF), China and Nepal, 2016
4. Landslide (rockfall) hazard in Upper Barun Valley, Nepal, 2017.

The frequent natural disasters in the Himalayan region points to the fact that the region is highly unpredictable and hence vulnerable to such threats. As a result, the hydropower advancements in the region get seriously hampered and investors are apprehensive to put their money on such risky projects.

6 Result and Discussion

6.1 Possible Solutions to the SHP Problems in India

Supplying electricity to remote villages (which are still devoid of electricity) through a state/national grid is economically not viable. This is due to the fact that these remote villages are many and far away from each other and each village has a small population of approximately a few dozens to a few hundred. Due to this, heavy transmission and distribution costs to each of these villages catering to only a few hundred people in every village at most are difficult to justify from financial point of view. The solution to this problem is a decentralized generation and distribution of electricity in such remote places. Electricity can be centrally generated with available resources here and then be distributed with a local grid to every household in the area. In IHR states, SHPs can be the source of generation. This is called as captive power plant i.e. electricity generation facility managed by an industrial or commercial energy user for their energy consumption. The benefit of such captive power plants is that they can operate off-grid but if the need arises due to excess generation, can be connected to the grid to avoid wastage of generated energy. Hence, where ever possible, captive users of SHP should be promoted. Central financial assistance and guidance are available for people who are willing to set up such captive power plants. Such schemes should be advertised on a mass scale to encourage people to become captive users. In this age, when it comes to electricity, people in such remote places should become pro-active and self-reliant and not just fully depend on the government to do the needful. Although this is easier said than done considering that people living in such remote places are generally from economically weaker sections of society, it is also true that with the right guidance and attitude, such villages can be electrified which will cause a paradigm shift in the lives of these people and their coming generations.

The first stage in setting a SHP plant is site selection. Proper planning has to be done for site selection as it is the most crucial part of the whole project as the site cannot be changed after the completion of the project. Since there is a constant threat of natural hazards in the state, proper and in depth risk mapping and hazard susceptibility must be done before starting the project. It is important to do a topographical survey (contour view of the site) which will give us a clear idea of the entire area and therefore enable us to take appropriate decisions on discharge calculation and capacity of the catchment area [13]. In addition to this, the latest technology must be used in forecasting and early warning systems. Emergency action and response plan must be in place in case of any eventuality along with a disaster preparedness and management system. Also, a detailed study must be done on the environmental and socio-economic impact of the SHP plant. Only if the site is favorable in the above-mentioned parameters the project should go ahead. After the site is locked, planning of design, operation, and optimization takes place.

The next stage in setting up an SHP plant is getting the legal compliances. Since many different clearances have to be taken at various levels, this process usually becomes a tiring, time-consuming, and daunting task. To make this process easier and user friendly, the government must introduce a 'single window fast track clearance system' for such SHP projects. Such an initiative will boost investor confidence and help the commissioning of the project much earlier. Also, this would increase the ease of doing business ranking of the state as well.

The next and probably the most important step consist of raising the capital for the project. There exist a few schemes by the government which provide financial assistance to people for setting up such plants. Efforts must be done to popularize such schemes as many interested people are unaware of such schemes by the government. Along with interest free or very low-interest loans, technical end-to-end support must also be provided for the smooth implementation and functioning of the project. For corporate investors who wish to set up SHP plants for commercial purposes, tax rebates in profits can be provided at least in the first few years. This will make SHP a lucrative and potentially profitable business venture for investors.

One of the major costs in hydro projects is the construction of dams or reservoirs which regulate water flow and help in flood control and also provide freshwater for agriculture in addition to generating power. This cost can be avoided for SHPs especially micro and pico by using the run-of-the-river (ROR) hydroelectric generation method where little or no water storage is provided and as such construction of dam or large reservoir is not necessary. However, in some cases, a small storage reservoir can be used which is called a pondage. Another advantage of ROR hydro-electric power is that it eliminates emissions of carbon dioxide and methane gases caused due to decomposition of organic matter in the reservoir of a conventional hydroelectric dam.

The high initial cost of conventional hydro turbines is also the main hindrance in implementing SHP schemes. The cost of these plants can be brought down by using a Pump as Turbine (PAT) [14]. Also, the diversion system has to be planned i.e. whether an open system or closed system will be used to divert water from the source and transport it to the turbine [15]. Also, filters should be used before the inlet valve in case there is a high amount of silt or muck in the flowing water to minimize the damage to the turbine blades.

Another hindrance in SHP is the lack of qualified technicians for its repair and maintenance purposes. Most people who are technically skilled are reluctant to move to remote locations permanently. To solve this problem, the local population needs to be given hands-on skills training in SHP through short-term courses. This will immensely help the local youths to become technically skilled and employable elsewhere too.

Lastly, lack of motorable roads is a major bottleneck in any infrastructure development project due to difficulty in transporting equipment and man power in the proposed site. The same geography which provides the state with natural ascent necessary for SHP also causes a hindrance in the construction of motorable roads due to its difficult and uneven terrain. Although rapid development has been witnessed in road construction in the IHR region in the last decade, the problem still persists in certain remote locations. Completion of the roads will strengthen the last mile connectivity of every village with mainstream cities and towns which will, in turn, usher fast socio-economic growth of people residing in such far-off places.

Due to inherent risks and high costs involved in large and medium hydropower projects [16], it is more logical to develop and utilize small hydropower potential in India. If a large hydropower plant has to be shut down due to technical problems or for repair and maintenance work, then electricity supply is affected for a large number of consumers. However, if micro and pico power plants are installed in large numbers, with each supplying power to a small group of consumers, this problem can be addressed to a large extent. At the simplest and most cost-effective level, micro and pico hydropower plants can generate enough power to charge batteries which can then be used for running electrical appliances of low wattage especially LED bulbs [17], and provide lighting solutions to homes that are not yet electrified.

7 Conclusion

In this study, it was found that IHR states have vast potential for SHP but less than a quarter of that potential has been utilized to date. This shows that there is tremendous scope for growth in this sector in the region. Next, the various problems, risks, challenges, and bottlenecks associated with SHP were identified, both general and region-specific, which are the reason for the low implementation of such SHP projects. After a careful and in-depth assessment of these problems, an attempt has been made to provide possible solutions to every type of risk at every level.

Finding new potential sites in SHP and the risk assessment associated with it is an area of continuous study and research. This part has a lot of future scopes as more research is needed in various technical aspects of SHPs and their working. There are various types of turbines like Kaplan, Francis, and Pelton whose performances can be studied and the one best suited for a particular region can be adopted for that place. There is also vortex turbine that can be tested for SHP in the region. Another upcoming technology is the hydrokinetic turbine which can be set up in canals and controlled rivers.

Although SHP alone will not solve all the energy problems around the world, with the right planning and implementation it can complement the primary grid by serving as secondary power generating unit or decentralized electricity generation grid. This will make the grid more reliable and robust.

References

1. Renewables, market analysis and forecast from 2019 to 2024. Fuel report, Oct 2019. Available from: https://www.iea.org/reports/renewables-2019/power (2019)
2. Yuksel, I.: As a renewable energy hydropower for sustainable development in Turkey. Renew. Sustain. Energy Rev. **14**, 3213–3219 (2010)
3. Rojanamon, P., Chaisomphob, T., Bureekul, T.: Application of geographical information system to site selection of small run-of-river hydropower project by considering engineering/economic/environmental criteria and social impact. Renew. Sustain. Energy Rev. **13**, 2336–2348 (2009)
4. Nautiyal, H., Goel, V.: Sustainability assessment of hydropower projects. J. Clean. Prod. **265**, 121661 (2020)
5. Doso, O., Gao, S.: An overview of small hydro power development in India. AIMS Energy **8**(5), 896–917 (2020)
6. Dave, S.K., et al.: Small, mini, micro & pico power plant: scope, challenges & deployment in Indian context. IJAERD (2015). e-ISSN (O): 2348-4470
7. Ministry of Power, Government of India: https://powermin.gov.in/en/content/power-sector-glance-all-india
8. Annual Report MNRE 2020–21: https://mnre.gov.in/img/documents/uploads/file_f-1618564141288.pdf
9. Chhabra, N., Pandey, K.K.: Risk Identification & Taxonomy in Small Hydro Projects in Uttarakhand. ResearchGate (2015). ISBN: 978-81-924713-8-9
10. Kayastha, N., Giri, S.: Geomorphological disaster in Chamoli, Uttarakhand, India: a challenge of finding a balance between infrastructural development and social and economical protection. ResearchGate (2021)
11. Thayyen, R.J., et al.: Hanging glacier avalanche (Raunthigad-Rishiganga) and debris flow disaster of 7th February 2021, Uttarakhand, India, a preliminary assessment
12. Pandey, P., Chauhan, P., et al.: Cause and process mechanism of rockslide triggered flood event in Rishiganga and Dhauliganga river valleys, Chamoli, Uttarakhand, India using satellite remote sensing and in situ observations. J. Indian Soc. Remote Sens. (2021)
13. Palakshappa, K., et al.: Feasibility study on proposed micro hydroelectric plant at Kappadi (Byndoor), Karnataka, India. ResearchGate (2019)
14. Motwani, K.H., et al.: Cost analysis of pump as turbine for pico hydropower plants—a case study. Procedia Eng. Elsevier. 1877-7058 (2012)
15. Anaza, S.O., et al.: Micro-hydro electric energy generation—an overview. Am. J. Eng. Res. (2017). e-ISSN: 2320-0847
16. Agarwal, S.S., Kansal, M.L.: Issues of hydropower development in Uttarakhand region of Indian Himalaya. ResearchGate (2017)
17. Yadav, G., Chauhan, A.K.: Design & development of pico micro hydro system by using house hold water supply. IJRET (2014). e-ISSN: 2319-1163

The Use of IoT to Improve the Energy Efficiency of the Fez Meknes Exhibition Hall

Reda Jabeur, Najat Ouaaline, Abouzaid Soufiane, Lakrim Abderrazak, and Houda Harkat

Abstract The mean goal of our current research work is to study the thermal optimization of the Exhibition Hall in an existing structure in Fez, which is considered a vital issue in morocco. The external walls and roofs will be insulated, and an efficient air conditioning, cooling, and heating system will be installed. The building envelope is simulated with TRNSYS as a multi-zone building by employing a specific flexible occupancy scenario, which allowed a proposal of solutions for thermal insulation and the integration of intelligent technologies, such as the Internet of Things (IoT). As an efficient technology for connecting objects, the Internet of Things (IoT) enables better exploitation and collection of building data. This is the ability to connect objects (sensors and actuators) by interacting them with their physical environment, allowing for better control of energy consumption patterns of active and passive customers. Therefore, using dynamic thermal simulation tools such as Revit-Insight 360 and TRNSYS is the main objective to provide a comparative assessment of several possibilities and solutions available to reduce the energy consumption of the exhibition hall.

Keywords Energy efficiency improvement · Thermal simulation · Heating load · Cooling load · IoT

R. Jabeur (✉) · N. Ouaaline · A. Soufiane
Hassan 1 University, Settat, Morocco
e-mail: r.jabeur@uhp.ac.com

N. Ouaaline
e-mail: Najat.ouaaline@uhp.ac.com

L. Abderrazak · H. Harkat
Sidi Mohammed Ben Abdellah University, Fez, Morocco
e-mail: abderrazak.lakrim@usmba.ac.ma

H. Harkat
e-mail: Houda.HARKAT@usmba.ac.ma

M. Bendaoud et al. (eds.), *The Proceedings of the International Conference on Electrical Systems & Automation*, https://doi.org/10.1007/978-981-19-0035-8_11

1 Introduction

Morocco's two primary energy policy challenges are a shortage of exploited natural resources and the enforcement of energy efficiency regulations. The energy consumption of our country is increasing; it increased from 0.34 TOE/capita in 2003 to 0.57 TOE/capita in 2018 [1]. Local primary energy production provided only 10.2% of the country's needs in 2018, and the rest is met by imports [3].

The construction sector is was found to be the largest energy consumer, accounting for more than 25% [4] of the whole consumption. Residential buildings consume 72% [5] of overall building energy, with the rest going to tertiary buildings (government, schools, hotels, hospitals, etc.).

Therefore, the building sector represents an enormous savings opportunity; about 40% can be saved by integrating several elements (efficient lighting, attractive surroundings, and well-maintained equipment).

In order to solve some issues related to energy consumption, the kingdom of morocco adopted several strategic steps in relation to this problem. Since 2008, daylight saving time (GMT + 1) has been an effective technique to mitigate the impact of overlapping usage, particularly during peak hours.

In 2014, this institutional initiative saved about 29 GWh of energy [6]. Implementing a new law, RTCM (Thermal Construction Regulation in Morocco), has allowed conserving about 1.3 Mtoe in 2020. The objective of the first iteration of this rule is to increase the energy efficiency of the building envelope in the new construction by using efficient insulation and glazing systems.

The intelligent building management system is responsible for monitoring, controlling, and optimizing building services (heating, cooling, visualization, etc.). It goes without saying that increased comfort in the indoor environment means increased energy consumption. This is a challenge for intelligent and efficient buildings to balance occupant comfort and energy consumption [7].

By introducing a lot of gadgets that are compatible with specific components of the home technology system. These technologies are based on Wi-Fi data transport, making it possible to have practically the house at your fingertips without the need for additional wiring or equipment modifications; this is also called the IoT (Internet of Things) technology.

The objective of this work is to increase both the summer and winter thermal comfort at the building level by parameterizing passive elements such as glass, building orientation, ventilation, and thermo-physical properties of materials [8, 9], while taking into account external variables (temperature, lighting, etc.), furthermore, by deploying a smart sensor that allows the collection and analysis of data on these many variables.

2 Building Description

2.1 Building External Views

The Chamber of Commerce, Industry, and Services of the Fez-Meknes region is currently working in partnership with the Ministry of Industry, the Fez-Meknes Regional Council, and the Fez City Council, the exhibition center project on the road to Sefrou. The location of this project is given in Fig. 1.

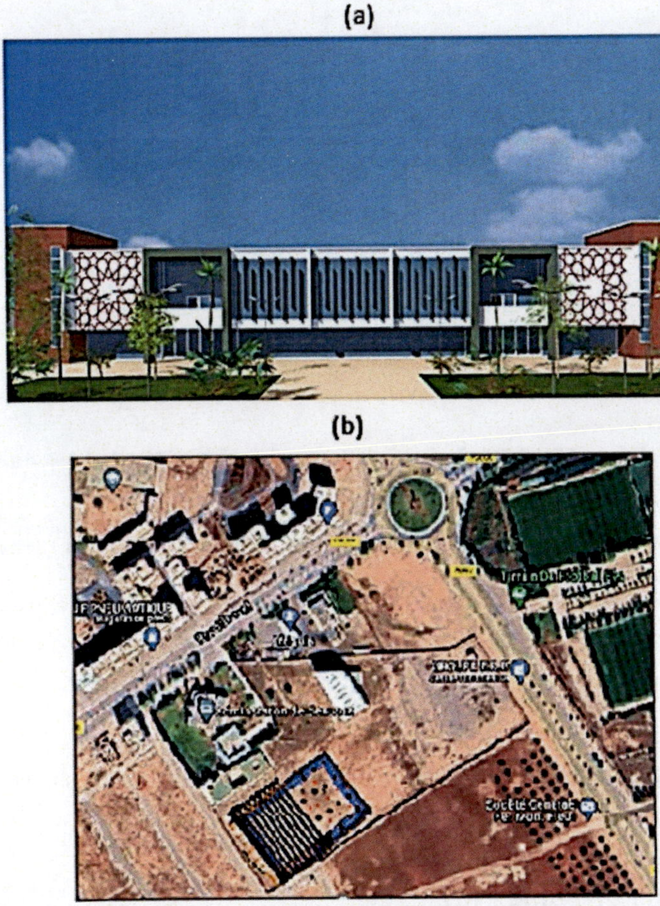

Fig. 1 **a** The computerized model of an architectural structure concerning its surroundings. **b** A Google maps satellite picture is depicting the project's location

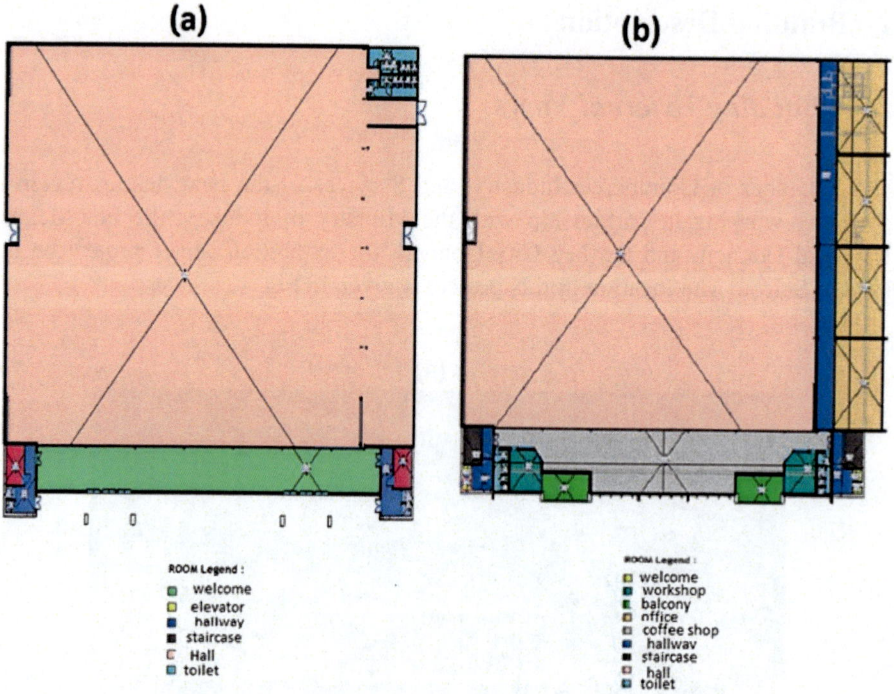

Fig. 2 Exhibition hall plan: **a** the ground floor, **b** the mezzanine

2.2 Exhibition Hall Plan

The exhibition hall is generally made up of a ground floor and a mezzanine, as illustrated in Fig. 2.

2.3 Building 3D Model

The modeling of the studied building (3D digital model) in Revit is shown in Fig. 3.

2.4 Internet of Things IoT

Internet of Things (IoT) devices are becoming increasingly popular to make buildings smarter and more energy-efficient. As buildings consume a huge amount of energy, smart buildings are intended to measure, limit, and control their energy consumption without compromising occupant comfort or operational efficiency [10]. Lighting,

Fig. 3 D model of the studied building using Revit software

electrical outlets, heating, ventilation, and air conditioning systems in buildings are large energy consumers.

The application of various IoT sensors in these systems helps buildings to become more intelligent and flexible. These IoT sensors and actuators create huge data (big data). Building analytics can be used to retrieve, filter, analyze, and make use of the huge amounts of IoT-enabled smart building data. Through the use of big data analytics, building energy efficiency and the overall user experience can be monitored and improved [11] (Fig. 4).

The development of the IoT guarantees positive energy gains in terms of consumption while ensuring comfort inside the building. Thus the easy integration of new energy sources, including renewable energy (solar energy).

2.5 *Thermal Procedure Simulation*

In order to study the thermal behavior thoroughly of this building, depending on several characteristics (location, building materials, overall architecture, energy concept chosen, etc.). The Trnsys software is used to integrate all the characteristics of a building and its equipment (heating and cooling systems). The simulation diagram of this building from Trnsys Simulation Studio is presented in Fig. 5.

Insight enables architects and engineers to design more energy-efficient buildings with advanced simulation engines and to build performance analysis data integrated into Revit. The building performance analysis team at Autodesk Company had a large number of projects in Autodesk Beta that made their way into Revit; sometimes, it seemed like they were made up of a lot of separate pieces.

These days, Insight 360 is a comprehensive and consistent guide to increasing the energy and environmental performance of buildings.

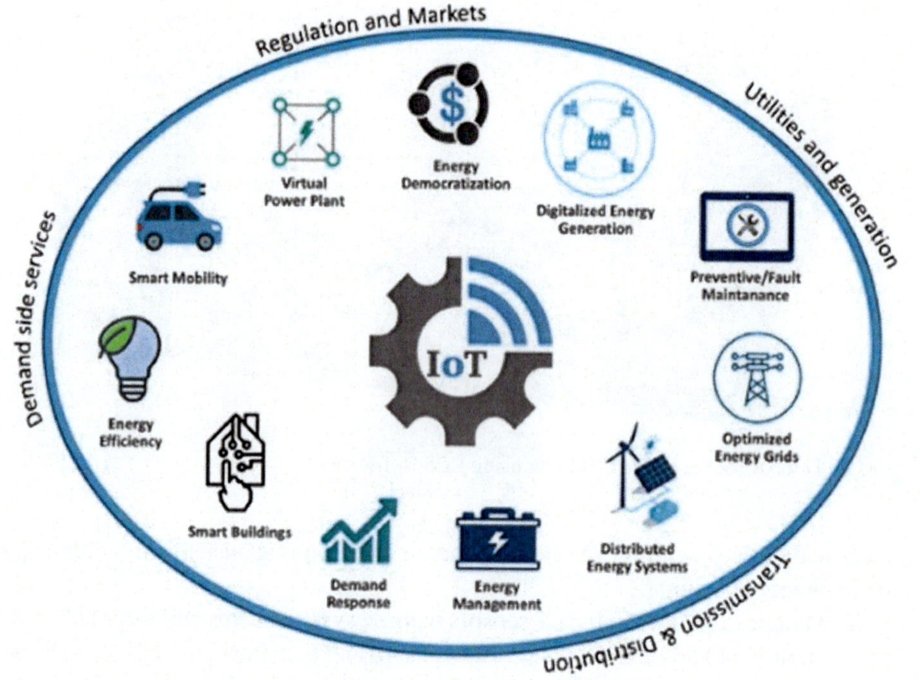

Fig. 4 Applications of IoT in an integrated smart energy system [12]

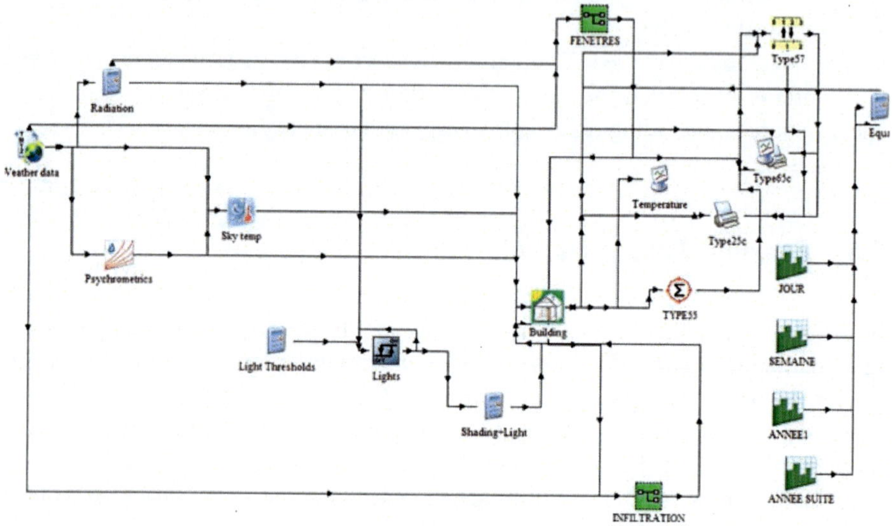

Fig. 5 Simulation diagram from Trnsys Simulation Studio

3 Automation System Description

Summer and winter set point temperatures are set by the automation system to ensure not only thermal comfort but also to control and manage the various electrical loads and monitor the energy used by the building's equipment.

The PLC is the brain of this automation system, which communicates with all of the sensors and actuators. This pre-programmed to achieve the ideal situations and specifications suited to our requirements, the PLC is an industrial controller that serves to command and operates many devices at once autonomously without human interaction.

Sensors and actuators are connected to the PLC inputs/outputs, and if necessary, extension modules are added. Communication protocols (Modbus, Profibus, Profinet) are used to communicate the PLC with its environment.

Switches and push buttons (PLC digital inputs) are used by TOR modules to control lamps (PLC digital outputs). As the temperature varies in an analog manner, analog inputs/outputs are used to control HVAC systems with PID controllers.

Air-Wall Stone Exchanger (also known as Moroccan well and more recently as climate well) is an exchanger that uses extremely little energy while cooling or heating the vented air in a building. This helps our new exchanger system work properly. Figure 6 shows this sort of heat exchanger in use in a passive dwelling.

To manage the whole system control and supervision, we used four touch screens are used and connected to each other as well as to PLC through ethernet protocol. Figure 7 describes the automation system.

4 Results and Discussions

The input/output values (analog) variation allowed the building comfort to be easily achieved with minimal energy consumption (Fig. 8).

4.1 Comparison of Energy Demand Results Between TRNSYS Software and Insight Autodesk

The Insight platform provides the average value of the building's annual consumption per m^2. Insight also gives several possible solutions to select the most optimal one.

In our case, we obtained the following energy consumption result: 218 kWh/m^2 per year.

Figure 9 shows an overview of the Insight platform, where we got our results and the other proposed solutions in order to select the best one.

Table 1 shows the simulation comparison results between Trnsys and Insight 360.

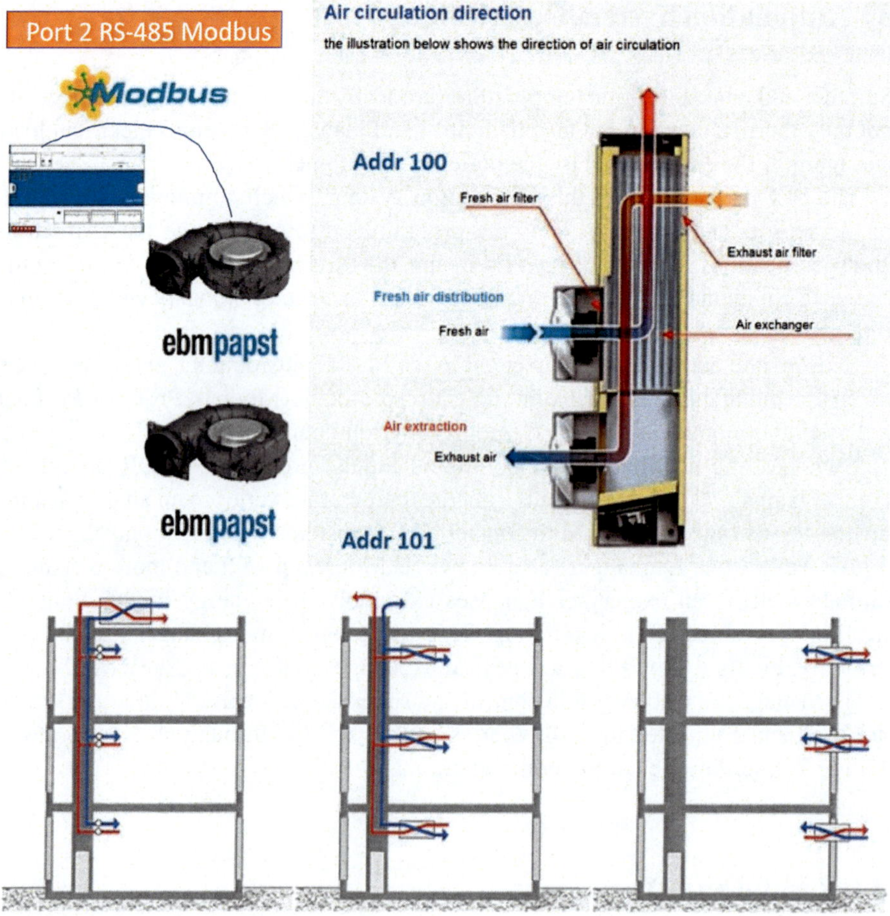

Fig. 6 Air exchanger—stonewall

Simulation results with Trnsys and Insight 360 give almost identical values. The absolute error obtained is 1.3% ((221 − 218)/221 = 0.013). This means that the energy simulation of both tools is successful.

5 Conclusion

This paper is set at the energy optimization of the design phase of the International Exhibition Hall of a building located in Fez. We used Revit design software to create a 3D model of the building. We then converted it to the Insight Autodesk 360 software to investigate the effect of sunshine on the building's thermal profile.

In a subsequent step, we integrated the project's many scenarios and material classes into the Trnsys dynamic thermal simulation software allowing for the energy

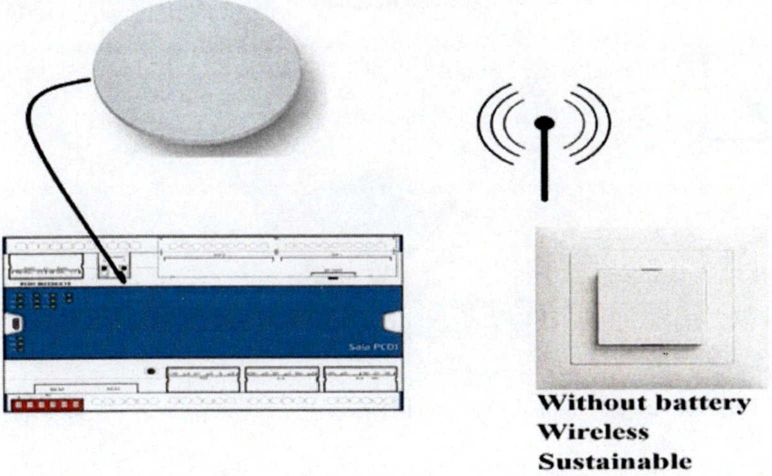

Fig. 7 Communication between the PLC and the wireless switch

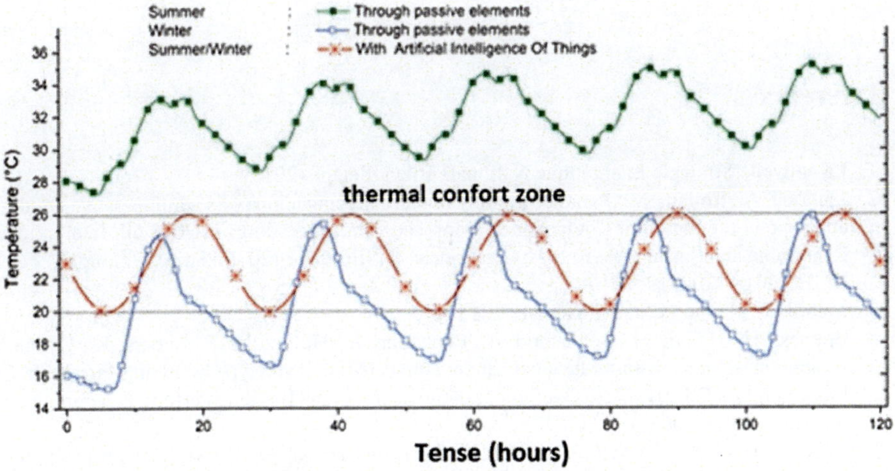

Fig. 8 Summer/winter temperature distribution profile in a building

simulation of the building for all four seasons of the year. Furthermore, Codes similar to the national building code should be adopted at the highest policy level for inclusion in building regulations. Modern technologies and secure platforms can be integrated with the internet of things to optimize energy use. Then the project's main objective is to manage and regulate energy resources in a cost-effective manner.

The study has shown how to achieve the comfort zone very efficiently without the use of active systems that generate energy losses, but rather through a low-energy-consumption ventilation system that compensates for these losses to ensure an energy balance through an intelligent energy management system known as the Internet of Things.

Fig. 9 Result of the analysis on the Insight platform

Table 1 Comparison results between Trnsys and Insight 360

Total of energy demand kWh/m^2/an	
Trnsys	Insight
221	218

References

1. La nouvelle Stratégie Energétique Nationale Bilan d'étape (2013)
2. Lamrani, A., Rougui, M.: Parametric study of thermal renovation of an administrative building envelope in the region of Chefchaouen (Morocco). In: Proceedings of 2018 6th International Renewable and Sustainable Energy Conference, IRSEC 2018 (2018). https://doi.org/10.1109/IRSEC.2018.8702919
3. Mobili, V.: Rapport d'activité annuel 2014
4. Bouhess, H., Hamdi, H., Benhamou, B., Bennouna, A., Hollmuller, P., Limam, K.: Dynamic simulation of an earth-to-air heat exchanger connected to a villa type house in Marrakech. In: Proceedings of BS 2013: 13th Conference of the International Building Performance Simulation Association, pp. 1063–1070 (2013)
5. Guechchati, M.A., Moussaoui, R.: Modélisation sous Trnsys du logement des professeurs de l'école Ain Elkhil a Ifrane. In: 1er Congres de l'Association Marocaine de Thermique, Settat, Morocco, 6–7 Mai 2010
6. Official journal of Morocco. N° 6360 (2014)
7. Stazi, F.: Energy efficient of building (2019)
8. Khattabi, M., Mharzi, M., Raefat, S., Meghari, Z.: On the application of a new thermal diagnostic model: the passive elements equivalent in terms of ventilation inside a room. IOP Conf. Ser. Mater. Sci. Eng. **353**, 012003 (2018)
9. Khattabi, M., Mharzi, M., Raefat, S., Garoum, M., Valančius, K., Meghari, Z.: A thermal diagnostic method based on a new approach of wall discretization in a dynamic state. J. Clean. Prod. **226**, 493–502 (2019)
10. Jia, M., Srinivasan, R.S., Raheem, A.A.: From occupancy to occupant behavior: an analytical survey of data acquisition technologies, modeling methodologies and simulation coupling mechanisms for building energy efficiency. Renew. Sustain. Energy Rev. **68**, 525–540 (2017)

11. Horban, V.: A multifaceted approach to smart energy city concept through using big data analytics. In: 2016 IEEE First International Conference on Data Stream Mining & Processing (DSMP), pp. 392–396. IEEE (2016)
12. Motlagh, N., Mohammadrezaei, M., Hunt, J., Zakeri, B.: Internet of Things (IoT) and the energy sector. Energies **13**(2), 494 (2020)

Use of Photovoltaic Energy in the Distillation and Purification of Water: Design of a Prototype

Kamilia Mounich, Aicha Wahabi, and Mohamed Chafi

Abstract Globally, the demand for good quality drinking water is growing. Indeed, the desalination and purification of sea water require lots of treatments. This paper presents the theoretical study and practical realization of a system that consists to solve the issue of water anti-calcification. Softening is the technique used to remove TH from water (due to the presence of alkaline earth salts: carbonates, sulfates, and chlorides of calcium and magnesium). This system will provide soft water, free of alkaline earth salts and heavy metals such as arsenic, lead, cadmium, nickel, mercury, chromium, cobalt, zinc, and selenium and which are highly toxic even in minor quantities. Hence, the principle of magnetic induction was studied as a key milestone to solve the calcite deposit into the canalization. This phenomenon creates strong magnetic fields that cause mineral ions (especially calcium carbonate) to agglomerate and neutralize with each other before the water is heated to reach then the final step which is the pass-through activated carbon filter.

Keywords Sea water distillation · Heavy metal · Water softener · Softener device · Limestone deposit · Anti limestone magnet

1 Introduction

Faced with a certain water crisis that is beginning to be felt and triggered the alarm all over the world, in addition to the economic constraints of sustainable development, appropriate solutions must be formulated to prepare for this challenge which threatens the survival of humanity. Morocco has considerable saline water resources and a large solar period throughout the year. It is necessary to use desalination technology by exploiting the available green energy [1]. This solution is a reliable way to produce drinking water. However, it should be noted that this method is economically reliable only for small units whose daily volume of drinking water varies from a few cubic meters to tens (or even hundreds) of drinking water per day [2].

K. Mounich (✉) · A. Wahabi · M. Chafi
LIPE, Higher School of Technology, Hassan II University, Casablanca, Morocco

M. Bendaoud et al. (eds.), *The Proceedings of the International Conference on Electrical Systems & Automation*, https://doi.org/10.1007/978-981-19-0035-8_12

The following study is focused on the practice of small processes. For large capacities, the combination of desalination processes with renewable energies requires high investment costs and the reliability of such systems is not always guaranteed [3].

The desalination process is divided into two main categories: firstly, the distillation processes (which require a phase change, evaporation/condensation) and secondly, the membrane processes (filtration), which consist to eliminate heavy Metals and bad taste from water thanks to filtration phenomenon [4, 5].

The best and most cost-effective method to ensure that is using a filter with multiple stages with activated carbon based as used in this experiment.

2 Process Working Principle

2.1 Materials and Methods

The study was conducted on a laboratory-scale distiller. It is consisting of a heat source (heating element) and a 750 ml container of sea water and a 6 mm diameter copper tube formed into 7 turns, a cooling fan powered through a 3.7 V lithium battery which in turn is charged through a 5 V, 12 W photovoltaic panel controlled by an SG09 servo motor, 4 LDR and an Arduino Nano Atmel Atmega 328p board. The freshwater passes through an anti-lime magnet to resolve the deposit of calcite in the pipeline and five steps activated carbon filter using gravity and absorption phenomenon (Fig. 1), and then to the final container to collect fresh water.

The five steps filter activated carbon-based offer a significant result in terms of water purification such as removing heavy metals, bad taste, and other deposits. This filter is a combination of five materials: Activated carbon, Anthracite, Sand, Garnet, and Gravet. Activated carbon and the Anthracite filters use minimal bonding agents and have superior contaminant reduction and performance capability.

Also, sand filtration technology can homogenize wastewater containing heavy metals. Thanks to the continuous flushing system, the resistance and flow rate of the filter bed are always stable and self-disciplined, providing stable output and steady-state operation. However, the Gravel layer was found to be very effective for the removal of sediment and heavy metals, even as the system clogged over time. In addition, the pores of the Garnet are small enough to allow the passage of liquid but too small to allow the passage of certain contaminating particles, which are filtered from the fluid.

2.2 Process and Theoretical Measures

The synoptic diagram (Fig. 2) describes the operating principle of the desalination system equipped, on the one hand, with a mechanism which supplies thermal energy

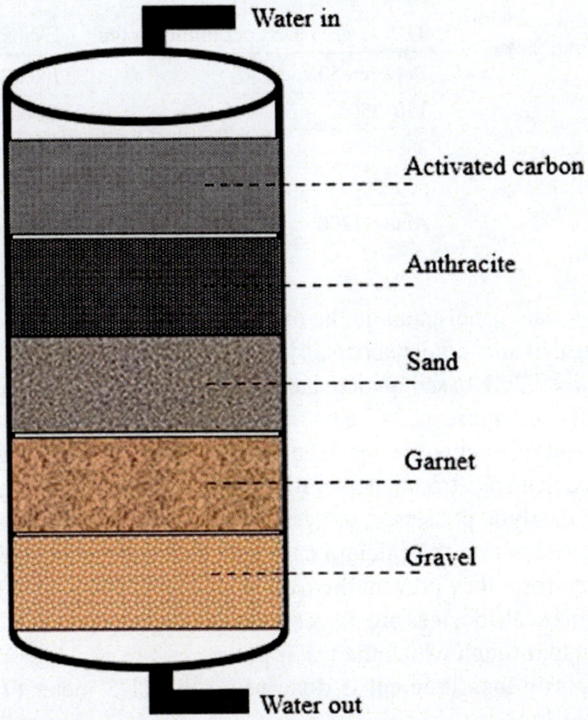

Fig. 1 Multiple steps filter (activated carbon-based)

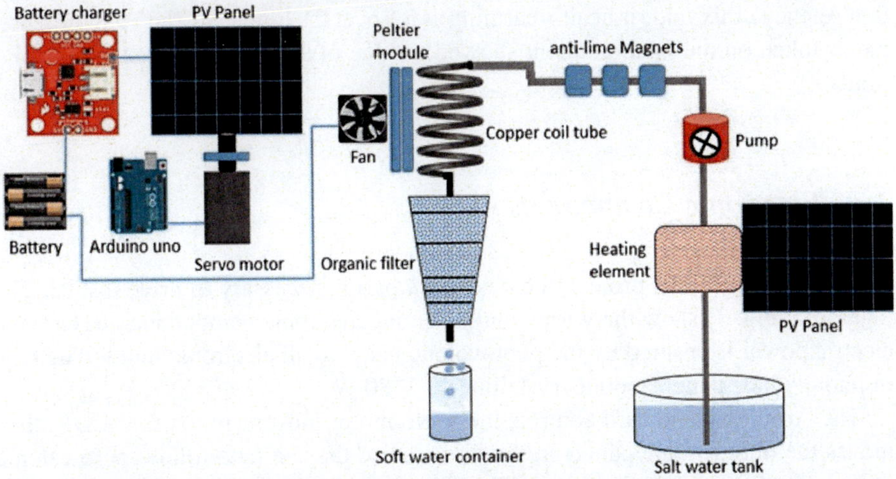

Fig. 2 Sea water distiller, purifying schema

Table 1 Taste of water with different concentrations of TDS

TDS level in parts per million (ppm)	Evaluation
Between 50 and 150	Excellent for drinking
150–250	Good
250–300	Fair
300–500	Poor
Above 1200	Unacceptable

through photovoltaic panels, and on the other hand, with a sub-system composed of two parts: the use of anti-lime magnets [4] to eliminate lime deposits in the pipes and the use of 5-stage filters based on activated carbon and other absorbent materials in order to eliminate the presence of heavy metals [5].

The permanent magnet processes generate a constant magnetic field: They are to be distinguished from electromagnetic processes which are at the origin of a variable magnetic field, catalytic processes, or systems at the origin of turbulent regimes.

These processes act on the calcium carbonate and facilitate its crystallization in a non-encrusting form they prevent the formation of scale.

Magnetic lime scale devices are placed in series on the pipe; the devices generate the magnetic field through which the water passes.

The water purity measurement is done using the TDS meter (Total Dissolved Solid) [6] (Table 1).

The experience showed a value of 10,600 ppm in saline water sampling measured by TDS before starting the heating. Hence, according to EPA (Environmental Protection Agency), this value is health-harming as it's over the limits. So, the TDS measurements token on the final recipient showed a value of 90 ppm, which is within EPA limits.

2.3 Electronic Components

In order to enable this project, the electronic part is necessary to drive results. The diagrams (Fig. 3) show the wiring and essential electronic components. In fact, the electric power is ensured by the photovoltaic panel to all electronic items. The type of photovoltaic panel is monocrystalline 12 V 80 W.

The Arduino Nano card controls the servo motor movement via four LDR, then moves the photovoltaic panels in the direction of the sun (sun follower) to catch a maximum of solar rays. The battery 7.4 V is under charge permanently to ensure the cooling system fan power supply. A Peltier module was added to guarantee a perfect cooling of the solenoid copper tube.

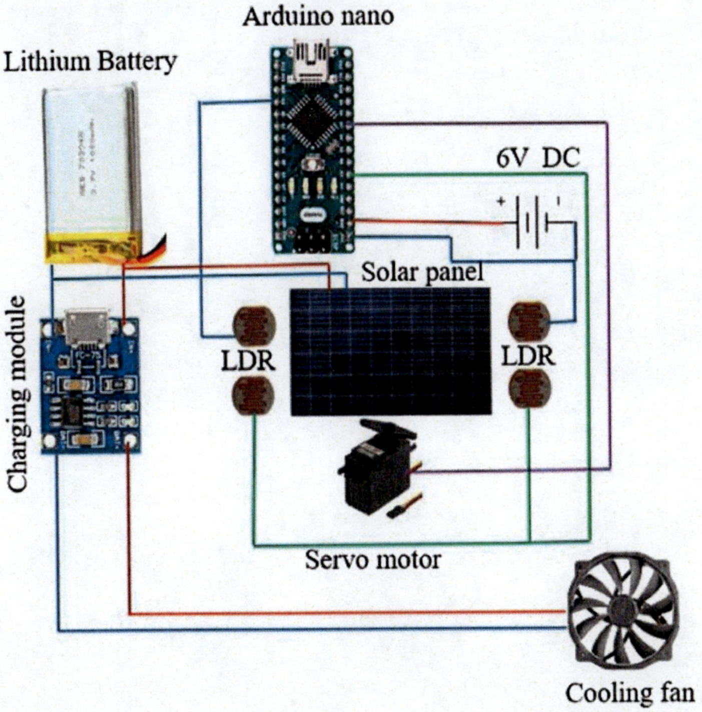

Fig. 3 Electronics schema

2.4 *Magnetic Anti-lime Scale Device*

Revolutionary anti-lime scale and anti-magnetic scale solution works without chemical additives, without maintenance, and current. This very powerful magnetic field prevents the formation of limestone and other deposits and cleanses pipes and equipment from old limestone encrustation.

The presence of limestone nanocrystals suspended in the water causes indeed a natural phenomenon of limestone erosion. The water that has become depleted in available calcium and carbonate ions begins to behave like rainwater: it seeks to become loaded with mineral salts. In other words, the principle is simple and effective; these magnets create a magnetic field that modifies the polarization of the mineral salts suspended in the water, thus preventing the formation of scale inside the water pipes (Fig. 4).

2.5 *Grouping of Photovoltaic Cells*

We have coupled two photovoltaic panels in parallel. A parallel mounting of two cells increases the generator output current. In a group of identical cells connected

Fig. 4 Anti-lime magnet (calcium build up)

in parallel, the cells are subjected to the same voltage and the characteristic resulting from the grouping is obtained by adding the currents (Fig. 5). Equation (1) in turn summarizes the electrical characteristics of a parallel association of NP cells.

$$V_1 = V_N \text{ and } I_T = I_1 + I_N \text{ (}N\text{is PV cells number)}$$

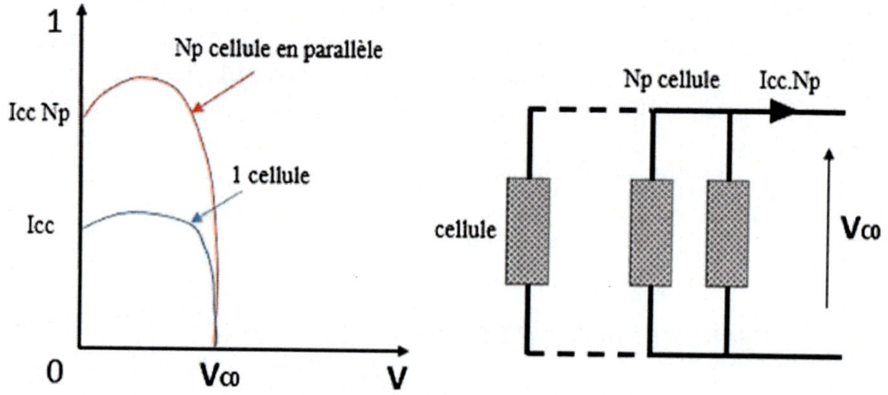

Fig. 5 Characteristic result from a grouping of Np cells in parallel

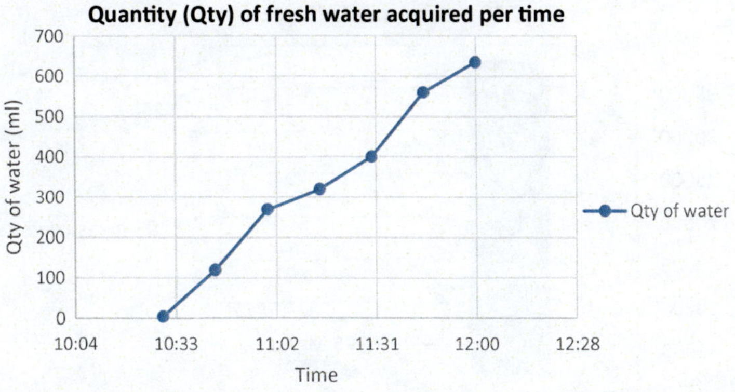

Fig. 6 Quantity of fresh water (ml) acquired per time

$$I_T = I \text{ total} \tag{1}$$

3 Results and Discussion

The day of September 9th, 2021, was chosen to perform the experience, which is characterized by strong sunshine throughout the day with a maximum temperature around 33 °C, specify that the time of the first drop was detected at 10:30 a.m. and the total amount of distillate collected until 12 p.m. is equal to 645 ml. The time between the first drop and the total quantity of collected fresh water is around 120 min (Fig. 6).

As mentioned before, the first fresh drop was collected at 10:28 even the condensation begin earlier as the water take several times to pass through the five stages of the filter. As the process progress, the quantity of water increase to reach finally around 640 ml at mid-day.

Among the results done, the biochemical test of the fresh water coming from the filter with the TDS device, and we had in Fig. 7.

4 Conclusion

It is well known that the scarcity of water is an object of both nature matter and human responsibility. The huge need for fresh water requires the exploitation of all available green energy such as solar energy, and at the same time take benefit that nature offers in terms of organic filters to provide fresh drinkable water for all needy areas. Indeed, this will not only gives alternatives but also will stop harming our planet by using fossil fuel.

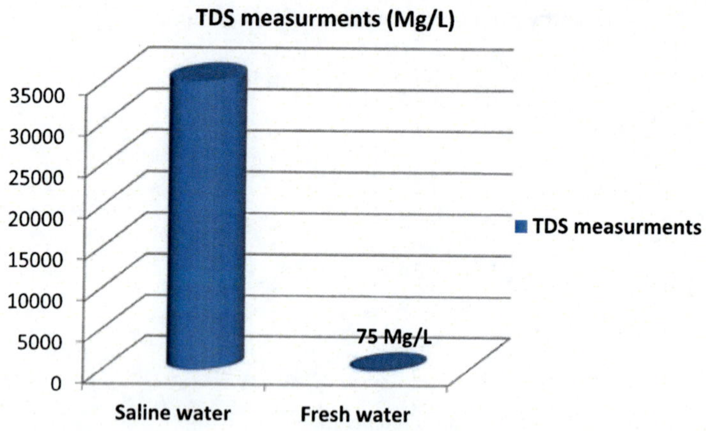

Fig. 7 Salt measurement TDS

Electromagnetic and magnetic techniques are a good solution to decrease the rate of heavy metals treatments because these non-invasive and cheap techniques do not need chemicals added to the water. There are various theories about the effects of these techniques. Other papers affirm that an electromagnetic field favors the formation of crystals within the fluid instead of in the pipes, and for this reason, the scale would not form. These techniques can increase or decrease turbulence in the fluid, promoting the aggregation or disaggregation of ferromagnetic and diamagnetic colloids. In addition, studies conducted at the University of Alicante have seen an orientation of the crystals formed after the application of an electromagnetic treatment that did not occur without treatment. It has also been observed in other studies that after the application of the magnetic treatment, the diffusion coefficients of the sodium, magnesium, and calcium ions increased and those of the anions decreased [7].

The filtration and absorption system provides a wide utilization in wastewater treatment. Nowadays, many prototypes of filters are available, competitive cost, easy to build, and efficient. As this research showed the benefits of a mixture of all absorbents material (Activated carbon, Anthracite, sand, Garnet, and Gravel), lots of the study demonstrated the strong influence of additions on the removal of heavy metals from saline water by using 5 layers filter, which varied according to the type of heavy metal.

The anti-lime magnet is fixed permanently in the water pipe, which gives autonomy and performance to the system. However, the treatments of different types of wastewater present a challenge. In fact, all types of water treatment will not be covered by a single filter, which gives the perspective to work on the filter change over or develop a filter that gathers all the characteristics that allow a standard filtration.

References

1. Mounich, K., Wahabi, A., Chafi, M.: Water-purifying distiller with a cooling system controlled by a photovoltaic panel. In: Motahhir, S., Bossoufi, B. (eds.) Digital Technologies and Applications. ICDTA 2021. Lecture Notes in Networks and Systems, vol. 211. Springer, Cham (2021)
2. Carr, W.J., Levin, B.H., Dissinger, M.L.: Water drinking and air drinking: some physiological determinants. Psychon. Sci. **13**, 23–24 (1968)
3. Esmaeilion, F.: Hybrid renewable energy systems for desalination. Appl. Water Sci. **10**, 84 (2020)
4. Delyannis, A.A., Delyannis, E.A.: Distillation processes. In: O. Sauerstoff. Oxygen (System-Nr. 3). Springer, Berlin, Heidelberg (1974)
5. Rajczykowski, K., Loska, K.: Stimulation of heavy metal adsorption process by using a strong magnetic field. Water Air Soil Pollut. **229**, 20 (2018)
6. TDS and PH meter: https://www.safewater.org/french-fact-sheets/2017/2/9/tds-ph. Accessed 2021/11/01
7. Martínez Moya, S., Boluda Botella, N.: Review of techniques to reduce and prevent carbonate scale. Prospecting in water treatment by magnetism and electromagnetism. Water **13**, 2365 (2021)

Towards FPGA Implementation of an Intelligent Hybrid Energy Management System

Asmae Chakir, Badr Chegari, Mohamed Tabaa, and Emmanuel Simeu

Abstract In this paper, we propose a state machine energy management system towards a software and subsequently hardware implementation embedded on FPGA (Field Programmable Gate Array) to plan the missions of the energy management state machine. The state machine is based on 8-active states. They are illustrated as follows. Two states for battery charging and discharging management during energy shortage. One state representing energy consumption from the grid and one state for hybrid consumption from PV and wind turbine. Two states for the direct consumption from the wind turbine and the injection of the PV production to the grid with and without battery charging. Finally, two states for direct consumption from PV and injection of wind generation to the grid with and without battery charging. To ensure the proper operation of the hardware pre-implementation, the state machine has been turned to a VHDL (VHSIC Hardware Description Language) code following the ISE Design Suite synthesis software. The results obtained show the efficiency of the state machine developed to complete the control states by meeting the transition constraints.

Keywords Hybrid renewable system · Energy management system · State machine · VHDL coding · Hardware/software implementation · FPGA

A. Chakir (✉) · M. Tabaa
Pluridisciplinary Research and Innovation Laboratory (LPRI), Moroccan School of Engineering Sciences (EMSI), Casablanca, Morocco

M. Tabaa
e-mail: m.tabaa@emsi.ma

B. Chegari
Research Foundation for Development and Innovation in Science and Engineering (FRDISI), Casablanca, Morocco

Engineering Research Laboratory (LRI), System Architecture Team (EAS), National and High School of Electricity and Mechanic (ENSEM), Hassan II University, 8118 Casablanca, Morocco

B. Chegari · E. Simeu
TIMA, CNRS, Grenoble INP (Institute of Engineering Université Grenoble Alpes), Université Grenoble Alpes, 38000 Grenoble, France
e-mail: emmanuel.simeu@univ-grenoble-alpes.fr

M. Bendaoud et al. (eds.), *The Proceedings of the International Conference on Electrical Systems & Automation*, https://doi.org/10.1007/978-981-19-0035-8_13

1 Introduction

The energy management concept started to be built since 1982 [1], as a part of power control that allows a system to interact easily with other entities belonging to the same electrical grid. The aim is to manage and control the energy consumption of buildings, industries, companies, factories, or others according to a specific control system. At present, the residential sector is considered the primary consumer of electricity. Therefore, more than 31% of the work involved in energy management systems improvement has been focused on residential buildings [2]. The use of energy management has also become increasingly important with the integration of renewable energies in the power generation sector. This is due to intermittency periods of renewable production that require energy efficiency.

With the smart grid concept and the implementation of communication and information infrastructures with a bi-directional aspect, advanced metering, energy storage technologies and renewable generation systems, the paradigm revolution in the use of electricity, conservation and energy optimization remains crucial [3, 4]. This revolution has greatly interested researchers in the electrical field, as they have tried to solve the energy management issue using several intelligent and/or linear methods. Whether it is for isolated applications or grid-connected [5].

Several studies have tackled the problem of energy management, the difference between them has been noticed on the energy sources used, the considered storage system, and the connection or not to the utility grid, otherwise the used energy management method [6]. In fact, the authors in [7] proposed a stochastic mixed-integer nonlinear programming to manage the energy of a micro-grid consisting of a non-renewable generation, PV plant, storage systems, an electric vehicle, and electric loads. A stochastic optimization approach was reported in [8] to schedule energy inside a smart household to facilitate the implementation of demand response programs. Similarly, based on mixed integral linear programming authors in [9] proposed a system to intelligently manage the energy of a smart home in order to ensure minimal energy costs while increasing the comfort of the consumer.

In [10], a real-time control was conducted using Dspace rapid prototyping system in order to balance at each instant the energy produced by PV-Wind hybrid system and the energy stored and consumed by loads. A nature-inspired method has been handled in [11] to develop a new meta-heuristic optimization for smart home appliances scheduling. The authors of [12] have developed five different operation modes of the PV-Wind-Battery hybrid system using a finite state schematic model; the transitions were performed by a nonlinear switching system. A control strategy is implemented in [13] with a fuzzy logic controller to smooth the power fluctuation of a PV-Wind hybrid system and to maintain the battery state of charge (SOC) within the allowed limits.

We note that most of the studies in the literature have used linear, dynamic, or predictive programming; otherwise, the energy management can be performed with a reduced state sequence algorithm [14]. Moreover, the linear programming solution is efficient to obtain an optimal solution according to constraints satisfaction for linear problems. However, it is not very efficient for most applications. However, non-linear programming is able to solve problems that are more complex even though it requires a very high number of iterations. For the same degree of program complexity, dynamic programming divides the optimization into sub-systems. All these methods are characterized by the speed of convergence, which can be solved by artificial intelligence, as well as by particle swarm optimization or genetic algorithms, although they are under the constraints of the parameter definition and arrest criteria [15]. This makes their hardware implementation difficult to handle.

In fact, the energy management of an electrical system must be done in a fully automatic way. The tasks for an autonomous system are considered very complex. In order to perform them with high performance, developers use hardware boosters [16]. There are several types of hardware accelerators, namely: GPU (Graphics Processing Unit), ASIC (application-specific integrated circuit) and FPGA (field-programmable gate array). It should be noted that FPGAs have been used according to the literature as they have the advantage of flexibility and being reconfigurable.

Indeed, it is difficult to ensure just with software implementation the real time reaction character for an energy management algorithm, especially adapted to a system of fluctuating character as the renewable production systems. In this sense, the authors on [17] studied all the dynamics via a hardware implementation of a hybrid Wind-supercapacitor-Battery system feeding a variable DC load. The result showed the real time dynamics of the micro-grid even with fluctuating load or wind conditions. This was done via an FPGA implementation. Also, a sub-model based on a Marcov model have been implemented on FPGA by the authors of [18] for the energy management of electric vehicles taking into account significant criteria, namely: the driving style and the planned trip. An experiment was conducted by the authors [19] to validate the management algorithm they developed for the management of a PV-battery-fuel cell based system. The FPGA was used in this case as an intermediate interface.

From the mentioned literature, few researchers have dealt with the case of the physical implementation. Nevertheless, it is a very important criterion for the solution selection and obviously, it depends on the strategy pursued accordingly. Indeed, this is the objective of our current paper. In fact, we aim to develop the first phase of the system implementation, which is based on the energy management algorithm adaptation to its test platform. For this, we have adapted an energy management algorithm for the PV-Wind-Battery hybrid renewable system [20] to a state machine that was then implemented on the Xilinx ISE (Integrated Synthesis Environment) synthesizer for a later FPGA implementation.

To this end, in the following section, we will start to present in detail the considered hybridization architecture. Then we will present our state machine energy management system. Section 4 is devoted to the simulation study and discussions. Otherwise, the conclusion and future work are drawn up towards the end.

2 Hybridization Strategy: Architecture

In Morocco, the solar and wind potential is considered as significant [21]. These two sources represent complementary energy supply systems [22]. Their use in hybrid energy system represents a cost-effective solution [23]. Such a system has proven its efficiency in a weather condition like the Mediterranean one. Besides, in most cases, to increase the reliability of a renewable system, a storage system is necessary. For this reason, we considered the PV-Wind-Battery system as an optimal architecture for an alternative renewable generation system [24].

For countries that have already been connected to the grid, such as Morocco, these systems may be connected to the grid to further benefit from the renewable character of these systems as we studied in [25, 26]. Besides, the characteristics and the passive part of the house held in view in this paper resulted from previous studies [27, 28]. Therefore, for our present case we propose the architecture presented on Fig. 1, which is a continuation of our previous work.

The architecture is based on a photovoltaic installation and a wind generation. These two sources are connected with DC/DC and AC/DC converters respectively. These two converters are used to extract maximum power from the renewable sources as well as to keep a continuous common bus voltage steady. The architecture represents a battery storage system with its bi-directional system to ensure both charge and discharge phases. The utility grid is linked via a bi-directional converter for the collaboration and injection stages.

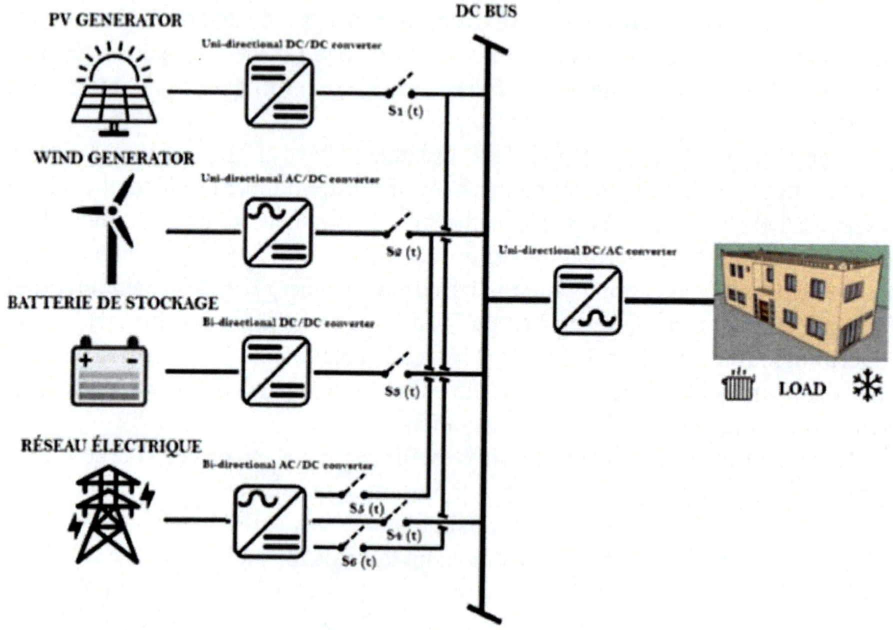

Fig. 1 Energy management system architecture

The six switches (S1, S2, S3, S4, S5 and S6) are used to facilitate the energy management of the hybrid system considered for our study. Below is the role of each switch.

- S1: This switch is used to directly transmit the energy generated by the PV array to the DC bus.
- S2: It is used to transmit directly the energy produced by the micro wind turbine to the DC bus.
- S3: This switch is used bi-directionally to initiate both the charge and discharge of the energy storage battery.
- S4: This switch is used to access the electrical grid to meet a possible energy deficit in the system.
- S5: It is used to inject all the energy produced by the photovoltaic field into the electrical grid.
- S6: This one is used to inject all the energy produced by the micro wind turbine into the electrical grid.

3 State Machine Energy Management System

Six switches will manage the hybrid system that their task has been well described during the previous section. These switches can be either open or closed so they represent a state of zero or one Eq. (1), respectively. According to the discrete probability distributions [29], the management system will be simulated at the flip of a coin six times. Therefore, we have 26 corresponding to 64 possibilities. However, the management system must meet physical constraints for an optimal control and implementation.

$$S_i \in \{0\} \cup \{1\}, i \in [\![1, 6]\!] \tag{1}$$

To ensure continuity of power consumption to the building's loads, the switches must not all be in low state. In return, for an optimal supply efficiency, the switches should not all be in high state. Which translates into Eqs. (2) and (3).

$$\sum_{1}^{6} S_i \neq 0 \tag{2}$$

$$\sum_{1}^{6} S_i \neq 6 \tag{3}$$

From the architecture shown in Fig. 1, we can deduce that switches S1 and S5 play a complementary role for photovoltaic generation. Moreover, the switches S2 and S6 play complements roles for the wind generation. Indeed, the system cannot ensure local production and at the same time inject electricity to the grid, Eqs. (4) and (5).

$$S_1 + S_5 = 1 \tag{4}$$

$$S_2 + S_6 = 1 \tag{5}$$

These constraints help us to eliminate 48 cases of switch states that cannot be produced for the proper operation of the management and control system. However, other states are to be eliminated because physically they do not represent tolerable states for the adequate performance of the architecture. Hereafter, the reasons for improving the states and the transition from 16 remaining states to only 8 states.

- 000111: To be eliminated because we cannot take energy from the network (S4 = 1) at the same time we inject renewable energy to the grid (S5 = S6 = 1).
- 001011: To be eliminated because we cannot take energy from the battery (S3 = 1), while the energy produced from the PV-Wind system is injected to the neighboring grid (S5 = S6 = 1).
- 001111: To be eliminated. Same reason as 000111.
- 010110: The system cannot take from the grid (S4 = 1), while it injects towards the grid solar energy (S5 = 1).
- 011110, 100101, 101101: Are to be eliminated for the same reason as 010110.
- 111100: It is not possible to consume from the grid and the battery and both renewable sources at the same time.
- 000011: Hybrid energy cannot be injected completely without consumption.

On Table 1, we have the states held in account for the generation of the energy management state machine of the PV-wind-battery hybrid system connected to the grid as represented in Fig. 2. Following are the notations for the transitions considered in the state machine.

$$x \Leftrightarrow P_{pv} > P_{load} \tag{6}$$

$$y \Leftrightarrow P_{wind} > P_{load} \tag{7}$$

Table 1 The eight remaining machine states

The 8 remaining states after updating under the form: [s1 s2 s3 s4 s5 s6]
010010
011010
100001
101001
110000
110100
111000 Battery discharge
111000 Battery charging

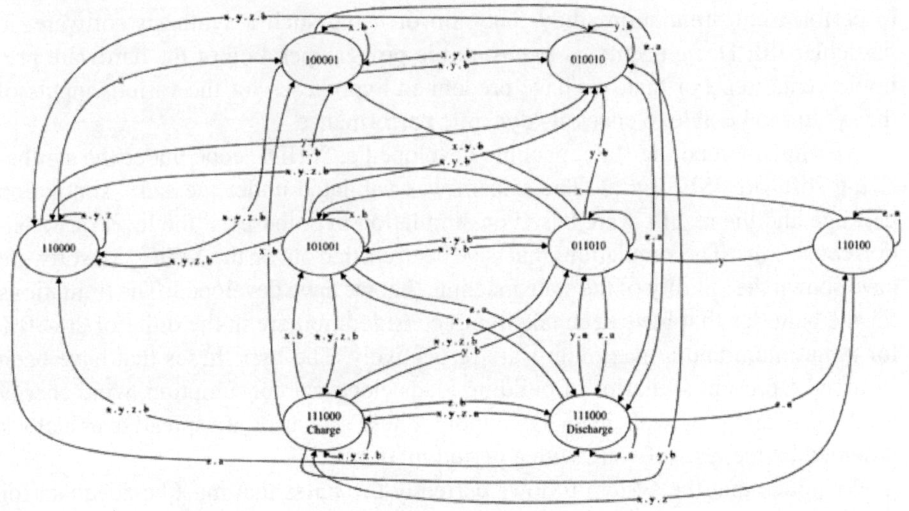

Fig. 2 Proposed energy management system state machine

$$z \Leftrightarrow P_{pv} + P_{wind} > P_{load} \tag{8}$$

$$a \Leftrightarrow SOC > SOC_{min} \tag{9}$$

$$b \Leftrightarrow SOC > SOC_{max} \tag{10}$$

where, P_{pv} is the energy produced by the photovoltaic panels, P_{wind} is the energy produced by the wind turbine. P_{load} is the energy consumed by the loads of the house. SOC is the state of charge and SOC_{min} is the minimal state of charge else SOC_{max} is maximal one. These powers and battery states of charge are considered as inputs for our control system and then the comparisons will allow us to see the states machine that are considered as output of our block.

4 Simulation Study

Our proposed solution is based on a state machine energy management system. Each state is characterized by a combination of the high and low (one or zero) states of the six switches that characterize the hybrid architecture. To move the system from state A to state B, the transition from A to B must be assured. According to the previous section, the transitions are based on the availability of renewable energy, the state of charge of the storage system, and the electrical consumption at a given time.

Before a power management system is actually implemented, it must go through a pre-implementation phase. To this end, we have chosen the Xilinx ISE software to perform our simulation study. The aim of using such a synthesis software, in particular ISE Design Suite, is to ensure the proper operation of the hardware pre-implementation. For that, we have predefined hypotheses for the various inputs of the system to be able to check its dynamic performance.

We implemented the state machine developed as VHDL code under the synthesizing software ISE, Fig. 3. The system was validated under the same simulation package and the results were based on simulation hypotheses at the level of transition constraints. The simulations that have been carried out on the Xilinx ISE software have shown the validity of the state machine that we have developed. The limitations for the batteries that have been taken into consideration are in the order of 20–80% for a minimum and a maximum state respectively. The hypotheses that have been assumed represent scenarios of building loads electrical consumption, wind energy availability and photovoltaic energy supply. These scenarios are spread over a clock provided by the test software with a period of 10 nF.

We notice that the system follows perfectly the states that must be adequate for each combination of the constraints presented by Eqs. (6)–(10). In fact, in case of solar potential availability that can cover totally the load of the house the state 100001 is activated (st1 between 0 and 24 nS), mentioned by st1 on the simulation, which makes the house consume all the power from the solar system. Depending on the hypothesis, this state is enabled during the first 10 nS. Moreover, what is produced by the wind system is injected in full into the neighborhood network. This is valid when the battery is on its maximum state; otherwise, the state 101001 is activated (st3 at 10 and 180 nS) so that the storage system takes the excess energy produced during this period. By the same process and assuming the coverage reliability of the building consumption by the wind energy, the state 010001 is activated (st2 between 24 and 54 nS, 104 and 134 nS, 184 and 214 nS). This occurs when the battery is in its maximum state of charge. If this is not the case, then the 011010 state is activated. When it is the role of the hybrid system to cover the electrical demand it is the 110000 state that is activated (st5 between 54 and 64 nS, 134 and 144 nS and 214 and 224 nS) when the battery is on its maximum state. Otherwise, if the battery is on its minimum state, the network comes in to cover the energy shortage during the concerned instant with the state 110100 (st6 between 74 and 94 nS, 154 and 174 nS). During the activation of the hybrid system, two cases are possible. If the hybrid energy is in excess and the state of charge of the battery is under its maximum state, then the 111000-state of charge is activated. Otherwise, if the hybrid system is not able to cover the energy demand at a given instant and the battery is above its minimum state, then the 111000-state of discharge is activated (st8 between 64 and 74 nS, 144 and 154 nS and 224 and 234 nS).

5 Conclusion and Future Work

This paper develops an energy management method for a grid-connected PV-Wind-Battery hybrid system based on a state machine. This latter consists of 8 states each

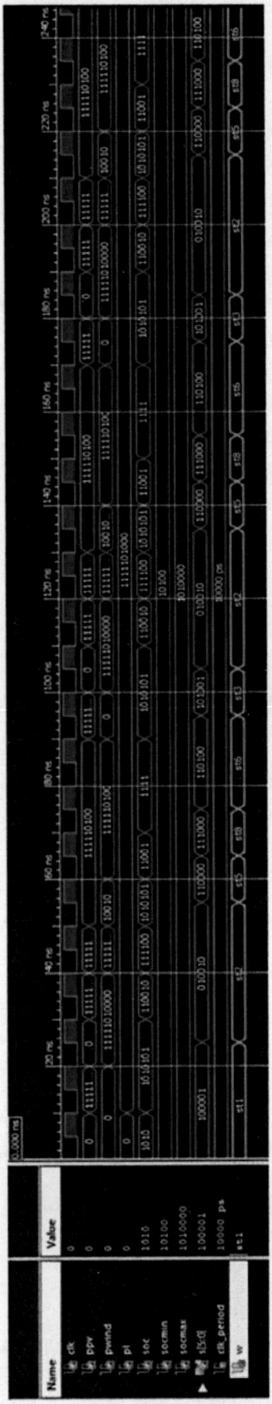

Fig. 3 Xilinx ISE simulation results

setting the six switches of the architecture according to a possible combination of two values taken by each S_i, namely: 0 or 1. The proposed state machine has been translated into a VHDL code and simulated on the Xilinx ISE synthesis platform. The transitions of the state machine have been taken as hypotheses or scenarios varying on a sinusoidal duration period, taken in this simulation as 20 nF. The simulations showed that the management system follows very well the adequate and optimal states for each instant t. However, this is only a test bench to validate the operation of the machine. While in the real case, we must first implement this system at the hardware level via the RTL diagram that we will have after synthesizing it on the software. In fact, The RTL code enters a phase known as logic synthesis, where the RTL code is translated into a set of logic components (adders, memory blocks, etc.) called a netlist. Then comes the physical implementation phase for an FPGA target. After that, couple the developed system with the environment generating the thermal loads of the building, especially those of the air conditioning and heating as well as the weather conditions for the generation of the hybrid system powers. This will allow us to effectively and physically test how our system can manage its energy flow in order to respond in an efficient way to the energy deficiency of this building.

References

1. Lee, D., Cheng, C.C.: Energy savings by energy management systems: a review. Renew. Sustain. Energy Rev. (2016)
2. Shaikh, P.H., Nor, N.B.M., Nallagownden, P., Elamvazuthi, I., Ibrahim, T.: A review on optimized control systems for building energy and comfort management of smart sustainable buildings. Renew. Sustain. Energy Rev. **34**, 409–429 (2014)
3. Zhou, B., Li, W., Chan, K.W., Cao, Y., Kuang, Y., Liu, X., Wang, X.: Smart home energy management systems: concept, configurations, and scheduling strategies. Renew. Sustain. Energy Rev. **61**, 30–40 (2016)
4. Azuatalam, D., Paridari, K., Ma, Y., Förstl, M., Chapman, A.C., Verbič, G.: Energy management of small-scale PV-battery systems: a systematic review considering practical implementation, computational requirements, quality of input data and battery degradation. Renew. Sustain. Energy Rev. **112**, 555–570 (2019)
5. Olatomiwa, L., Mekhilef, S., Ismail, M.S., Moghavvemi, M.: Energy management strategies in hybrid renewable energy systems: a review. Renew. Sustain. Energy Rev. **62**, 821–835 (2016)
6. Mariano-Hernández, D., Hernández-Callejo, L., Zorita-Lamadrid, A., Duque-Pérez, O., Santos García, F.: A review of strategies for building energy management system: model predictive control, demand side management, optimization, and fault detect & diagnosis. J. Build. Eng. **33** (2021)
7. Silva, J.A.A., López, J.C., Arias, N.B., Rider, M.J., da Silva, L.C.P.: An optimal stochastic energy management system for resilient microgrids. Appl. Energy **300**, 117435 (2021). https://doi.org/10.1016/J.APENERGY.2021.117435
8. Merdanoğlu, H., Yakıcı, E., Doğan, O., Duran, S., Karatas, M: Finding optimal schedules in a home energy management system. Elsevier (2020)
9. ur Rehman, U., Yaqoob, K., Adil Khan, M.: Optimal power management framework for smart homes using electric vehicles and energy storage. Int. J. Electr. Power Energy Syst. **134**, 107358 (2022). https://doi.org/10.1016/J.IJEPES.2021.107358

10. Kumar, P.S., Chandrasena, R.P.S., Ramu, V., Srinivas, G.N., Babu, K.V.S.M.: Energy management system for small scale hybrid wind solar battery based microgrid. IEEE Access **8**, 8336–8345 (2020). https://doi.org/10.1109/ACCESS.2020.2964052
11. Hussain, I., Ullah, M., Ullah, I., Bibi, A., Naeem, M., Singh, M., Singh, D.: Optimizing energy consumption in the home energy management system via a bio-inspired dragonfly algorithm and the genetic algorithm. Electron **9**, 406 (2020). https://doi.org/10.3390/ELECTRONICS9030406
12. Noghreian, E., Koofigar, H.R.: Power control of hybrid energy systems with renewable sources (wind-photovoltaic) using switched systems strategy. Sustain. Energy Grids Netw. (2019). https://doi.org/10.1016/j.segan.2019.100280
13. Das, S., Akella, A.K.: A control strategy for power management of an isolated micro hydro-PV-battery hybrid energy system. In: Proceedings of the 4th International Conference on Electrical Energy Systems. ICEES 2018, pp. 397–401 (2018). https://doi.org/10.1109/ICEES.2018.8442350
14. Ammari, C., Belatrache, D., Touhami, B., Makhloufi, S.: Sizing, optimization, control and energy management of hybrid renewable energy system—a review. Energy Built Environ. (2021). https://doi.org/10.1016/J.ENBENV.2021.04.002
15. Zia, M.F., Elbouchikhi, E., Benbouzid, M.: Microgrids energy management systems: a critical review on methods, solutions, and prospects. Appl. Energy (2018)
16. Hireche, C.: Etude et implémentation sur SoC-FPGA d'une méthode probabiliste pour le contrôle de mission de véhicule autonome (2019)
17. Kumar, V.N., Panda, G.: FPGA implementation of power management algorithm for wind energy storage system with Kalman filter MPPT technique. In: Proceedings of the IEEE International Conference on VLSI Design, Jan 2018, pp. 449–450 (2018). https://doi.org/10.1109/VLSID.2018.107
18. Mahmoud, D.G., Elkhouly, O.A., Azzazy, M., Alkady, G.I., Adly, I., Daoud, R.M., Amer, H.H., Elsayed, H., Guirguis, M., Abdelshafi, M.G.: Intelligent battery-aware energy management system for electric vehicles. In: IEEE International Conference on Emerging Technologies and Factory Automation ETFA, Sept 2019, pp. 1635–1638 (2019). https://doi.org/10.1109/ETFA.2019.8869496
19. Pati, A.K., Sahoo, N.C.: An experimental study on energy management of grid-connected hybrid PV-battery-fuel cell system. In: 2020 IEEE Calcutta Conference CALCON 2020—Proceedings, pp. 20–24 (2020). https://doi.org/10.1109/CALCON49167.2020.9106485
20. Chakir, A., Tabaa, M., Moutaouakkil, F., Medromi, H., Alami, K.: Optimal control design for a grid connected PV-wind-battery hybrid system feeding residential loads. In: ACM International Conference Proceeding Series (2019)
21. Meliani, M., El Barkany, A., El Abbassi, I., Moumen Darcherif, A., Mahmoudi, M.: Smart grid implementation in Morocco: case study. Mater. Today Proc. **45**, 7675–7679 (2021). https://doi.org/10.1016/J.MATPR.2021.03.176
22. Jurasz, J., Canales, F.A., Kies, A., Guezgouz, M., Beluco, A.: A review on the complementarity of renewable energy sources: concept, metrics, application and future research directions (2019). https://doi.org/10.1016/j.solener.2019.11.087
23. Weschenfelder, F., de Novaes Pires Leite, G., Araújo da Costa, A.C., de Castro Vilela, O., Ribeiro, C.M., Villa Ochoa, A.A., Araújo, A.M.: A review on the complementarity between grid-connected solar and wind power systems. J. Clean Prod. **257**, 120617 (2020). https://doi.org/10.1016/J.JCLEPRO.2020.120617
24. Chakir, A., Tabaa, M., Moutaouakkil, F., Medromi, H., Alami, K.: A combined source and demand-side energy management system for a grid-connected PV-wind hybrid system. In: Innovations in Smart Cities Applications, 3rd edn, pp. 707–721 (2020). https://doi.org/10.1007/978-3-030-37629-1_51
25. Chakir, A., Tabaa, M., Moutaouakkil, F., Medromi, H., Julien-Salame, M., Dandache, A., Alami, K.: Optimal energy management for a grid connected PV-battery system. Energy Rep. (2019). https://doi.org/10.1016/j.egyr.2019.10.040

26. Chakir, A., Tabaa, M., Moutaouakkil, F., Medromi, H., Alami, K.: Smart multi-level energy management algorithm for grid-connected hybrid renewable energy systems in a micro-grid context. J. Renew. Sustain. Energy **12**, 055301 (2020). https://doi.org/10.1063/5.0015639
27. Chegari, B., Tabaa, M., Simeu, E., Moutaouakkil, F., Medromi, H.: Multi-objective optimization of building energy performance and indoor thermal comfort by combining artificial neural networks and metaheuristic algorithms. Energy Build. **239**, 110839 (2021). https://doi.org/10.1016/J.ENBUILD.2021.110839
28. Chegari, B., Tabaa, M., Moutaouakkil, F., Simeu, E., Medromi, H.: Local energy self-sufficiency for passive buildings: case study of a typical Moroccan building. J. Build. Eng. **29**, 101164 (2020). https://doi.org/10.1016/J.JOBE.2019.101164
29. Sugiyama, M.: Examples of discrete probability distributions. In: Introduction to Statistical Machine Learning, pp. 25–36 (2016). https://doi.org/10.1016/B978-0-12-802121-7.00014-5